Dune Country

A naturalist's look at the plant life of Southwestern sand dunes

Dune Country

Janice Emily Bowers

Foreword by Ann Zwinger

Drawings by Margaret Kurzius

Photographs by Steven P. McLaughlin and Janice Emily Bowers

The University of Arizona Press / Tucson

For Steven, with love

This book features many of the dune fields in the American Southwest, summarizing morphology and vegetation and describing access. The information was up-to-date as of 1986, but travelers should be aware that accessibility in particular may have changed.

First University of Arizona Press edition 1998

∞ This book is printed on acid-free, archival-quality paper.
Manufactured in the United States of America
03 02 01 00 99 98 6 5 4 3 2 1

Library of Congress Cataloging-in-Publication Data
Bowers, Janice Emily.
[Seasons of the wind]
Dune country: a naturalist's look at the plant life of southwestern sand dunes / Janice Emily Bowers; foreword by Ann Zwinger; drawings by Margaret Kurzius; photographs by Steven P. McLaughlin and Janice Emily Bowers.–1st University of Arizona Press ed.
p. cm.
Originally published: Seasons of the wind. Flagstaff, Ariz.: Northland Press, c1986.
Includes bibliographical references and indexes.
ISBN 0-8165-1890-4 (alk. paper)
1. Sand dune plants—Southwestern States. 2. Sand dune plants—Mexico. 3. Sand dunes—Southwestern States. 4. Sand dunes—Mexico. I. Title.
QK142.B68 1998
581.7'583'0979–dc21 98-2957
CIP

British Library Cataloguing-in-Publication Data
A catalogue record for this book is available from the British Library.

Contents

Foreword

GIVEN THAT DUNES MAKE UP A SCANT ONE percent of the desert landscape, the power of their presence far outweighs the amount of land they cover. As Janice Bowers writes, when you say "desert" to someone, the response is invariably, "dune." That immediate association speaks to the potency of the image—an unseamed interface between sand and sky, an inviolate flowing horizon line, the perfect restful curve, the prompt psychological response to the angle of repose.

If I had a chance to choose someone with whom I'd enjoy wandering a dune field, Janice Bowers would be high on my list—for a dozen reasons. She is a well-qualifed working botanist and a competent outdoorswoman. She is one of those quietly enthusiastic companionable people who wears especially well in the out-of-doors because she is at home there. She is generous in sharing her knowledge. Her observations are those of a trained scientist and a perceptive woman who knows beauty when she sees it.

Like Janice Bowers, I've driven along a highway, seen a

dune presence rise on the horizon, and thrown away the rest of the schedule to go get sand in my shoes. There is something magnanimous and magnificent about a dune field. They call you from a long way off.

Wading across the Amaragosa River in California and looking up at the Dumont Dunes was to hear trumpets in golden sky. The Kelso Dunes moan, one of the few "singing" dunes in this country. I recently drove miles out of my way to find Sand Mountain, near Fallon, Nevada. I've had water nearly frozen overnight in my cup in the chill gypsum sands of White Sands National Monument.

I have even seen dunes from the air and so strong was their call that even though it took me a year to get there, I finally did. I first saw the dunes north of Winnemucca, obliterating the sagebrush flats, looking like (it seemed then) an inexplicable sand pile strewn by a careless hand. When I slogged out to them on the ground, it was as if the dunes managed to keep an even half-mile ahead, moving as I moved, stopping when I stopped.

When I finally reached the crests and hollows, the dunes were alive: laced with small plants that scrambled across the sand, in bloom with small butterflies and minute grasshoppers which popped ahead of my footsteps like drops of water on a hot skillet. The sand fumed off the crest of the barchan that towered above me and the whole dune field was animate and exhilarating and alive.

Janice Bowers has captured that animate quality of dunes for those of us who are dune freaks, who *like* having sand in our pockets. We can only rejoice when someone so skilled describes how plants adapt to constantly shifting sand, make the best of a tough situation. If you don't think dunes are always on the move, all you have to do to become a believer is to lie down on a dune on a windy day. Within a few minutes you're half drifted with sand. It's an instant lesson in respect for your plant counterparts blossoming a few inches away. The wind laid them there in the first place and the wind continues to mold those satisfying sinuous shapes.

The surprisingly kindly water regime of dunes enables them

to support a rather amazing flora. Still, the harsh conditions of dunes impose a severe series of adaptations upon those plants that would survive: they must be able to withstand aridity, quickly elongate their stems so that their flowering and leafing parts can operate above the sand. This stringent environment often produces endemism. Endemics are a delight anywhere, but more so on the dunes.

She evokes the stringency of the dune habitat but also the delicacy of what grows there, the ephemeral annuals, the few perennials: dune evening primrose, four delicate white petals cupping open in the evening; the delightful little sweet-pea flowered *Astragalus,* endemic to the Algodones Dunes; an ajo lily, translucent petals and sepals trumpeting in the spring.

The variety of dunes delights the imagination. Just in case there is anyone who thinks that if you've seen one dune you've seen them all, a reading of this book will dispense with that myth. With a lyrical turn of phrase, Janice Bowers describes the purity of shape, the dynamic aspects of color and movement. And there's an advantage to a book: you can go back and resavor the experience at will.

As much as those of us who love dunes will revel in this book, it is even more a book for all people who think that dunes are wastelands, a good place to dump trash or to chase around with dune buggies. Since they are not likely to read a book about dunes, perhaps each of us should gift one of our less dune-devoted fellow men with this book, for this kind of a book is the best way to make others aware of what a dune is: quietly, without polemics or shrill demands, just a relaxed conversation about a special part of the world a knowledgeable author finds infinitely fascinating.

This book could have no better fate than to attract and enchant the people who understand dunes least. More than ever we need to communicate the value of the quiet places, the worth of the empty space, the sense of the open horizon. The charm and scholarship of *Seasons of the Wind* is that it does just that.

All of us who love the abrasive quiet of a sand dune and the thunder of saltating sand grains, love the flow of contour and shifting profile of a dune, can be thankful that Janice Bowers has written with such exhilarating eloquence and such unmistakable understanding.

ANN ZWINGER
Colorado Springs, Colorado

Acknowledgments

WITHOUT THE RESEARCH OF MANY SCIENTISTS before me, I could not have written this book. I am also indebted to many friends for their help: Steven McLaughlin, best friend, companion, and colleague, who accompanied me in the field and prodded the book through many drafts; Tony Burgess, who read the manuscript at several stages and had many helpful comments and suggestions; Ray Turner, who made it possible for me to visit several dune fields in the Sonoran and Chihuahuan deserts; Margaret Kurzius and Ben Brown, who helped me explore the dunes near Desemboque; Elaine Cook, who gave form and direction to the project early on; Bob Webb, who made the manuscripts of his book on off-road vehicles and his paper on plant succession available to me; Gary Nabhan, who was the catalyst for the book; and Mark Taylor, who printed our photographs. Many people supplied information about sand dunes in the Southwest, including Thomas L. Jensen, Charles E. Collins Jr., J. Ross Arnold, James S. Morrison, Robert Sharp, Roger D. Zortman, Donald R. Harper, Bob Quesenberry, Jim Neal, Pete Eidenbach, and Julio Betancourt. I am grateful to all of them.

The knife-edge crest of the Kelso Dunes overlooks the valley floor five hundred feet below.

Introduction

THE HIGHEST SLOPES IN THE KELSO DUNE FIELD are so steep I must crawl up them on hands and knees. Hand over hand, plunging my arms wrist-deep into the sand, I pull myself up the dune as sand slides out from under me. My heart pounds. My panting drowns out the whistle of wind across my ears. As I reach the knife-edge crest that divides this mountain of sand in two, I rise slowly to my feet, knees trembling from the effort I have made. I feel that I am standing on the rim of the world; a single misstep and I will plummet off the edge. I take a few steps forward. My body is still shaking, and I lose my balance and fall. The sand catches me, and suddenly I realize there is no danger here unless the shock of beauty can kill.

Looking to the desert floor five hundred feet below, I see hills and hollows of sand like rising and falling waves. Now, at midmorning, they are paper white. At dawn they were fog gray. This evening they will be eggshell brown. The ridge where I stand rises steeply behind me to a conical peak and unwinds before me in a lazy S curve, half in bright sunlight, half still in shadow. I can hardly believe I am here, that I finally made it back.

I remember an afternoon several springs ago when my best friend and I stumbled upon these dunes by accident. We were on our way to someplace else, as usual, and had no time to stop and explore. Through the car window I saw a mountain of sculptured sand on the desert plain, dunes carved upon dunes, the highest peaks bathed in the golden light of late afternoon, the lowest ones already dimmed by shadow. As we drove by, I wondered if I would ever see the Kelso Dunes again.

Later that night I lay sleepless for hours, picturing the Kelso Dunes and other dune fields I had seen. I remembered the time I had first glimpsed the Death Valley Dunes from miles away in the Grapevine Mountains. The dunes were dwarfed by the distance before them and the mountains behind and seemed too small for the landscape. It was a gusty day, and wind had blown up sand and dust until the air was white. The Cottonwood Mountains behind the dunes were half-hidden in the haze; only the topmost peaks floated above the storm. As I drove closer to the dunes, ribbons of sand skittered across the road. Even inside the car, sand gritted between my teeth. As I parked and rolled down the window for a photograph, the car rocked in the wind. Low, rounded dunes near the car blended into conical peaks in the distance, their outlines blurred by the plumes of sand that fluttered from every crest. It was too windy to tour the dunes, or so I thought, and I drove on, watching them dwindle in my rearview mirror.

I remembered that many years before, I had seen White Sands, too, and recalled the blinding brightness of the sand and the way it crunched underfoot, more like snow than sand. The dunes seemed to roll to the horizon, each one a mirror image of the next. At the White Sands Visitor Center I learned that visitors may backpack into the heart of the dune field, and I wondered then why anyone would want to. When I saw the Kelso Dunes, though, I understood why.

What did these three dune fields have in common, I wondered? How did they differ? Were there others?

I set out to learn what I could about sand dune country in the Southwest, reading hundreds of papers by geologists, bot-

The dune field in Death Valley is dwarfed by mountains to the east and west.

anists, and other scientists, and traveling from California to Texas, from Utah to Coahuila, to see dunes for myself.

I learned that the Kelso Dunes, the Death Valley Dunes, and White Sands are just three dune fields among dozens scattered throughout the southwestern deserts, but that because the deserts are vast and the dune fields are few and far between, most people drive thousands of miles without ever seeing them; that dozens of species of plants grow just on sand dunes, nowhere else, and that certain lizards, snakes, insects, and rodents also live only on dunes; that the sand dune habitat poses special problems for plants that live there, and that sand dune plants have adapted to their habitat in surprising ways; and that some people see sand dunes in purely negative terms. One geologist even suggested that the best way to preserve deserts is to confine off-road vehicles to dunes, where they cannot hurt anything. After all, he implied, dunes are just wasteland, are they not?

Or are they? The more I learned about dunes, the more time I spent trudging across them at all seasons of the year, the less like wasteland they seemed.

The dune country is a lively place. Lizards, like clumps of animated sand, dart out from underfoot. Unseen rodents whistle piercing notes. Seed pods rustle across the sand. Drooping grass-leaves trace lazy circles in the sand like compass needles turned by the wind. The bathpowder scent of sand verbena drifts downwind. Sand is burning hot on one side of the dune, silky cool on the other.

Although sand dune ecosystems appear simple, they are actually complex and delicately balanced. Far from being wasteland fit only for a racetrack, sand dunes in the desert Southwest are unique, beautiful, fragile environments in which plants, animals, sand, sun, wind, and rain all fit together to make a dune country that is part of but separate from the surrounding desert.

This book is about the ecology of sand dunes, particularly of plants on dunes. The first chapter surveys the dune country in the Southwest and briefly reviews the four great North American deserts. The next chapter describes how dunes are formed; subsequent chapters detail the rigors of the sand dune environment and explain how some plants have adapted to them. Other chapters examine survival, succession, evolution, and distribution of sand dune plants. An appendix offers a selective guide to dune fields in the Southwest.

Although based in large part on research papers written for scientific journals, this book is a popular account of sand dunes and their plants, and I have tried to write the book using nontechnical language as much as possible. Unavoidable technical terms are defined in the text. In a effort to make the book less abstruse to readers who lack training in biology, I have used common names of plants throughout. I must confess that I am guilty of a cardinal botanical sin: that of inventing common names for some plants. In my defense, I can only say that most dune plants lack widely accepted common names and that I have used existing common names wherever possible. Scientific equivalents of common names used in the text are listed at the end of the book. Readers interested in technical accounts of the plant ecology and geology of sand dunes will find additional information in the chapter notes and the Selective Guide to Southwestern Dunes.

The Dune Country

ALTHOUGH MOST OF US MIGHT AUTOMATICALLY think of dunes when we think of deserts, dunes cover less than one percent of the North American deserts; and although we might think of deserts when we think of dunes, dunes are by no means confined to deserts. In Alaska, dunes perch on permanently frozen ground above the Arctic Circle; in northern Colorado, dunes are surrounded by fir and aspen forests; and about one-third of the Pacific Coast, from Cabo San Lucas at the southern tip of Baja California to Cape Flattery in northern Washington, is lined intermittently with dunes.

The setting for the dune country described in this book is the arid lands of the greater Southwest, the vast desert that stretches from southeastern California to western Texas and from central Utah to the Mexican states of Sonora, Chihuahua, and Coahuila. Biologists have divided this immense, arid region into four smaller deserts: the Chihuahuan, Great Basin, Sonoran, and Mojave. Each has a distinct assemblage of plants and animals, a characteristic climate, and some wonderfully interesting dune fields.

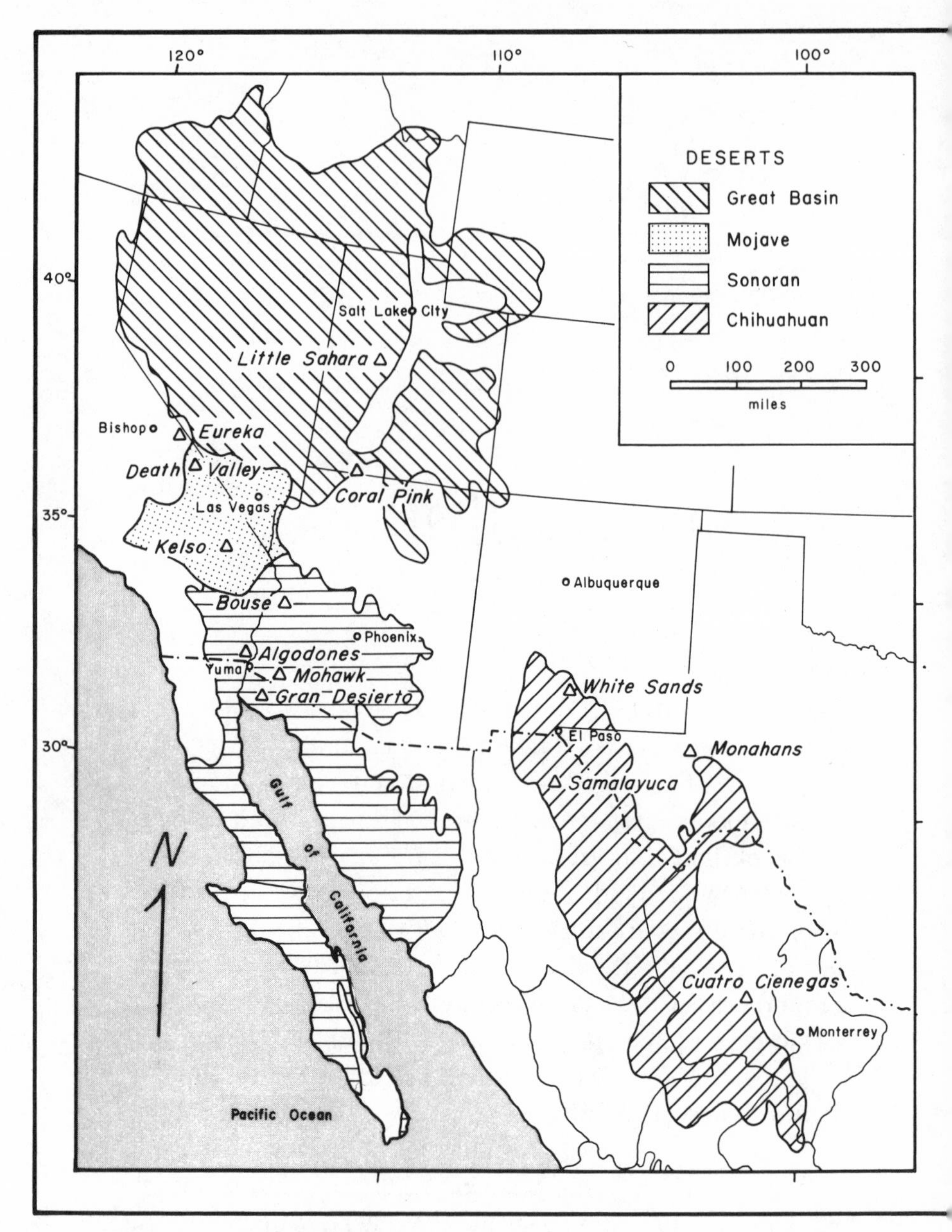

Southwestern Dune Fields

The largest desert in the Southwest is the Chihuahuan Desert, which covers over 200,000 square miles, largely in northeastern Mexico, but also in western Texas and southern New Mexico. The Chihuahuan Desert lies at moderately high elevations compared to other deserts: from 2,000 to 5,000 feet above sea level in its northern part and from 5,000 to 6,000 feet in the southern part. Winter temperatures are low, and freezing weather occasionally lasts for several days. Rain falls mostly in the summertime as local, drenching thunderstorms. Throughout much of the desert, runoff drains not into rivers, which eventually flow to the sea, but into basins ringed by hills and mountains. Characteristic Chihuahuan Desert plants are tarbush, creosote bush, whitethorn, marriola, candelilla, guayule, and lechuguilla. Many sand dunes in the Chihuahuan Desert were built with gypsum sand. White Sands, the most famous of these, is a national monument. Dunes in the Cuatro Cienegas Basin in Coahuila are also gypsum. Other Chihuahuan Desert dune fields are the Samalayuca Dunes and the Monahans Sandhills.

The Great Basin Desert is a close second in size to the Chihuahuan Desert and covers nearly 190,000 square miles. Much of Nevada and Utah are in the Great Basin Desert, as are parts of Oregon, Washington, Idaho, and Wyoming. Like much of the Chihuahuan Desert, the Great Basin Desert lacks outlets to the ocean. The desert floor is broken by hundreds of north-south-trending mountain ranges, the highest reaching 14,246 feet above sea level. Lying in the rain shadow of the Sierra Nevada–Cascade chain, the Great Basin Desert receives from four to twelve inches of precipitation annually, a little less than half of which falls in the summer. The Great Basin Desert is a cold desert. Winters always bring freezing temperatures and snow; the average temperature in January is only 28° F. Characteristic Great Basin Desert plants include big sagebrush, shadscale, rubber rabbitbrush, blackbrush, and greasewood. Much of the Great Basin Desert is sandy, with the sand well anchored by vegetation. One of the many dune fields in the Great Basin Desert is the Little Sahara Sand Dunes in central

Utah. The Coral Pink Dunes in southern Utah lie on the boundary between desert and woodland.

The Sonoran Desert covers about 120,000 square miles. Although much of it lies in northwestern Mexico, about half of Arizona and the southeastern corner of California are in the Sonoran Desert. The warmest of all the deserts, the Sonoran Desert rarely experiences freezing temperatures for more than twenty-four consecutive hours. Summer temperatures are often among the highest in the United States: daily highs of 120° F or more are not uncommon in the western half of the desert. Elevations range from below sea level in the Imperial Valley, California, to about 4,500 feet in mountain foothills. Like the Great Basin Desert, the Sonoran Desert is broken by mountains that stand like islands in a sea of desert. Unlike the Great Basin Desert, however, major rivers in the Sonoran Desert eventually reach the sea, draining into the Gulf of California. Precipitation ranges from less than two inches to about twelve inches annually, with a winter and a summer rainy season. Saguaro, blue palo verde, little-leaf palo verde, triangle-leaf bursage, ironwood, jumping cholla, teddy bear cholla, and creosote bush are some characteristic Sonoran Desert plants. Although there are fewer dunes in the Sonoran Desert than in the other deserts, the largest dune field in North America, the Gran Desierto Dunes, is in this desert. Other Sonoran Desert dune fields are the Algodones Dunes, the Mohawk Dunes, and dunes along the desert coasts of Sonora and Baja California in Mexico.

The Mojave Desert, which covers about 35,000 square miles, is the smallest North American desert and lies largely in southeastern California and southern Nevada. Elevations in the Mojave Desert range from 282 feet below sea level in Death Valley to about 4,000 feet above sea level in the foothills of isolated mountain ranges. Rainfall averages about three to four inches annually across the desert floor, increasing with rising elevation to about eleven inches. The hottest temperature ever recorded in the United States—134° F—was recorded in the Mojave Desert at Death Valley. During the summer, temperatures often exceed 100° F for one hundred consecutive days or more. Winters are colder than in the Sonoran Desert: in valley bottoms where cold air settles at night, temperatures drop below 0° F in winter and may be close to freezing even in the summer. White bursage, creosote bush, Mojave yucca, Joshua tree, Mojave sage, and Fremont dalea are some characteristic Mojave Desert plants. There are many small dune fields in the Mojave Desert, including the Eureka, Kelso, and Death Valley dune fields.

Four deserts and dozens of dune fields. The Eureka Dunes encompass only three square miles; the Algodones Dunes are more than sixty times as large. The White Sands dunes were built from grains of gypsum, the dunes in Death Valley from grains of quartz. In Saline Valley the dunes are no more than twenty feet tall; in the Eureka Valley they rise nearly seven hundred feet above the desert floor. The Gran Desierto Dunes may go for a year without any measurable rainfall; the Coral Pink Dunes are sifted over with snow every winter. Why are dune fields in the Southwest so diverse? How can sand dunes exist under such disparate conditions? The answers lie in the way dunes are created.

Of Wind and Sand, Sand and Rain

WE DROVE THROUGH KELSO LATE AT NIGHT, searching for the dirt road that would take us within a half-mile of the Kelso Dunes, marked on our map as the Devil's Playground. Gusts of wind buffeted the truck as we pulled to a stop. The dunes were a vague shape in the dark, ghostly pale and not quite real. We put up our tent as quickly as possible, but even with the tent, refuge eluded us that night. For hours we lay in our sleeping bags listening to the wind. Whirling through the pass between the Granite and Providence mountains, it roared like traffic on a distant freeway as it rushed toward the dunes. We braced ourselves as it shook the tent, tugged the shock-cords, flapped the rainfly. Gravel rasped across the ground-cloth, and sand sifted through the mosquito netting. Morning finally came but brought no respite from the wind, which still blew cold and steady from the southeast.

As we emerged from the tent, we saw that the sky was filled with sand and dust. Mountains to the north were enveloped in a gritty cloud, barely visible through the haze. In front of us was one of the four sand ranges that dominate the Kelso dune

Blowing sand on the Death Valley Dunes

field: massive, mountainous dune chains with peaks like glacial arêtes rising above long, sinuous ridges. Surrounding these sand mountains was a sand sea, a choppy sea awash with whitecaps. In the morning light we could see that the sand itself was neither white nor tan but some shade in between: the color of vanilla ice cream or of coffee lightened with milk.

After a gritty breakfast cooked on the tailgate in what little shelter the truck afforded, we plunged into the wind and headed for the dunes. With every step, clouds of sand swirled away from our feet. Sand beat against our flapping blue jeans, stung our bare hands and faces. A river of sand flowed around our ankles. Sand banners unfurled from the peaks and ridges of the highest dunes. The whole dune field was alive with rushing sand, a world in motion.

Sand beating against our hands and faces moved by saltation, a process in which individual grains of sand rise a foot or so into the air, fall, bounce into the air, fall again, bounce back up, and so on. The minimum wind speed at which saltation begins is between ten and seventeen miles per hour, depending on the size of the sand grains.

Sand flowing around our ankles moved by creep. Creep occurs at high wind speeds when the impact of grains bouncing on the sand during saltation drives other grains forward, bumping each other along the surface of the dune rather than blowing above it.

Wind whisks the finest grains of sand through the air in suspension. These settle to the ground as wind speed diminishes. Suspended sand at the Kelso Dunes had made the air white and obscured our view of distant mountains.

During saltation, sand flows up the gently sloping side of a dune, the windward side, and accumulates on the crest. When the crest is overburdened, grains tumble down the steep slope, also called the lee side or slip face, of the dune. Geologists refer to individual grains avalanching down the lee side as sand flow; they call the sliding of a cohesive mass of grains down the lee side slumping.

Sand flow and slumping make a dune inch forward as layer after layer of sand slides down its lee face. Saltation or creep of the windward slope keeps the rest of the dune from lagging far behind. Dunes move at varying rates, from one foot per year or less to between fifty and eighty feet per year. The rate of dune movement depends on the size and shape of a dune, the strength and direction of the prevailing winds, and the amount of vegetation on the dune.

To my untrained eye, the jumbled dunes in most fields seem to have no more organization or pattern than a plug of earth from an overturned flowerpot. Geologists, however, recognize several types of dunes, which can be categorized as active or stable dunes. Stable dunes, also called fixed dunes, are anchored by vegetation. Roots bind the sand and keep it from blowing away, while the leafy canopies of shrubs and grasses shelter the sand from direct blasts of wind. Stable dunes move only slightly, if at all. On active dunes, however, the plant cover is not dense enough to keep the sand in place. Many active dunes are barren where sand moves too freely and rapidly for

Tongues of sand show where slumping has occurred.

plants to prosper.

Barchans, which are crescent-shaped, are one type of active dune. The horns of the crescent point in the direction the dune is moving. Movement of barchans may be rapid, up to eighty feet per year. Barchan dunes are usually formed by winds that blow in the same direction most of the year. Transverse dunes, another type of active dune, are nearly straight sand ridges formed at right angles to the prevailing winds. Typically, transverse dunes dominate the center of most dune fields, whereas barchans are more common on the periphery. Star dunes, common in parts of the Middle East and North Africa, are rare in the greater Southwest, although there are star dunes in the Gran Desierto dune field in northwestern Mexico. Shaped like three- or four-armed starfish, these are active dunes created by winds blowing with nearly equal strength from several directions at different times of the year. Parabolic dunes are U- or V-shaped dunes that may be active but most

Wind creates ripples the same way it creates dunes.

often are fixed. The arms of parabolic dunes are anchored by plants, and only the noses creep forward.

Three conditions must be met for dunes to form: a source of sand, a source of wind, and a place for sand to collect. The sand source for the Kelso Dunes is a broad, sandy apron formed by the Mohave River as it flows out of Afton Canyon in the Cady Mountains west of the dune field. The river deposits an abundant load of sand at the mouth of the canyon during occasional floods. The source of wind for the Kelso Dunes is the prevailing westerlies, which funnel river sand to the east, forming a thirty-five-mile-long tongue of sand that ends at the Kelso Dunes. The place for sand to collect is a cul-de-sac made by the Granite and Providence mountains, which lie south and east of the dune field. Although winds blow more frequently from the west, strong winds buffet the Kelso Dunes from several directions throughout the year. Changes in wind direction confine the dunes, keeping them from engulfing the lower slopes of the nearby mountains or from spreading out in a sandy sheet across the entire valley. Such wind reversals also reverse the lee slopes, which face opposite directions at different times,

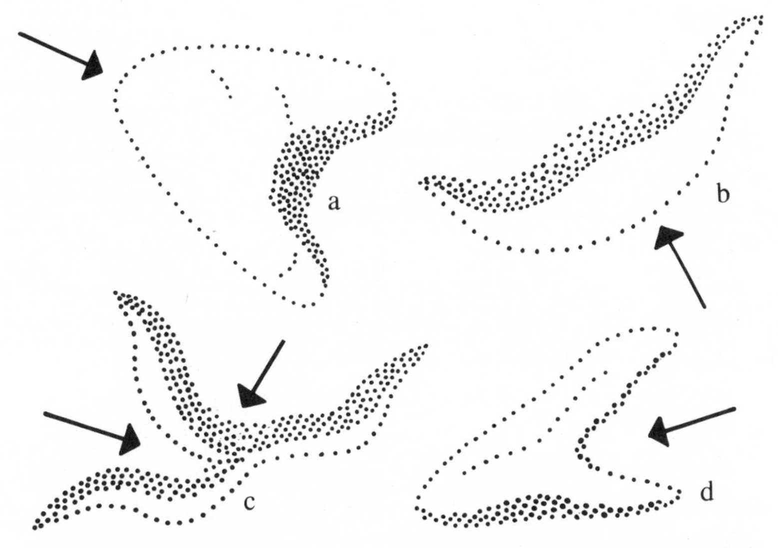

Types of dunes in the Southwest: a) barchan dune; b) transverse dune; c) star dune; d) parabolic dune. Arrows show direction of wind.

depending on how the wind has been blowing.

The same forces that shaped the Kelso Dunes also molded other dune fields in the Southwest. The source of sand differed from one dune field to the next, the prevailing winds might have blown from the east, north, or south, the place where sand collected might have been a terrace instead of a valley, but without an abundant supply of loose sand, without wind to lift and move the grains, without some topographic barrier to force the wind to drop its burden, no dunes would have been created.

Among sage-stippled hills lie the Little Sahara Dunes, as convoluted as crumpled paper, as wrinkled as a brain. The dune field, which covers three hundred square miles, is largely active, its white sand molded into ridges and furrows thirty to forty feet high or deep, depending on one's perspective.

We set up camp on the edge of the dune field, then relaxed in our folding chairs and watched a summer storm building to the south, where gray-blue thunderheads towered above the desert floor. A curtain of rain hid distant mountains, and gusts of wind brought the smell of dampened earth, the smell that means rain to all southwesterners. A tumbleweed bounded by us, then another and another. As a swirling dust devil enveloped us in grit and leaves and shreds of paper, we grabbed our chairs and ducked into the tent, not a moment too soon. The storm was upon us almost immediately. Rain pounded on the tent and was soon sloshing off the roof. The tent billowed as it filled with wind. There seemed to be a distinct possibility that we would become airborne, and we wondered if six plastic pegs were sufficient anchor for a 750-cubic-foot balloon. We concentrated so fiercely on the wind that we did not notice the water until it was too late. Runoff from the tent had collected on the hard-packed soil underneath, turning the floor of the tent into a waterbed. Our sleeping bags were soaked, but there was little we could do until the storm subsided, so we sat down to wait it out.

Eventually the worst of the rain and wind died down, and we ventured into the drizzle to see what could be done. With mixed interest and chagrin we noticed that although puddles had collected in the road and under our tent, the dunes themselves were merely damp. The surface of the sand was moist and pockmarked from raindrops, but showed no other sign of the rainstorm just past. This was, indeed, a vivid object lesson in the absorbing capacities of sand dunes.

Sand dunes are less arid than they appear; in fact, they are among the moistest of desert habitats. Like a sponge, dunes soak up every drop of rain that falls on them. No water runs off dunes, and since the top several inches of dry sand insulate moist sand beneath, little water evaporates. Even during hot, rainless summers, when the air temperature is 100° F and the sand surface is 140° F, the sand is cool and moist just six to twelve inches below the surface.

Not only do dunes lose less water than nearby desert flats, they provide more water for plants, as well. This is due to the different texture of desert and dune soils. Many nondune desert soils are mostly clay. Clay particles pack tightly in soil, creating innumerable tiny pores between the particles. Water accumulates in these pores, and because there is so much pore space, about twenty-five percent of a clay soil can be water. In clay soils, most of the water molecules are tightly held by the soil particles, and only free water—that is, water not adhering to soil particles—is available to plants. The remainder is held so tightly that even when the soil appears moist, plants can extract little water from it.

Sand grains are much larger than clay particles, and in sandy soils, the total pore space is much less than in clay soils. Even when sand is saturated, less than ten percent of the soil is water. Because water molecules are bound much less tightly by sand grains than by clay particles, plants can pull water more readily from sand or sandy soil than from clay soils, particularly when there is little water in the soil to start with. Since sandy soils have less pore space than clay soils, rainfall penetrates to greater depths in sand than in clay, providing extra moisture for deeply rooted sand dune plants.

Water is crucial to the survival of desert plants. If plants can find all the water they need on active dunes, sand dunes should be an ideal habitat for all desert plants. But this is not necessarily true, as barren dunes attest.

Survival

THE MOON WAS FULL, A CIRCLE OF DIMPLED FOIL pasted on a cardboard sky. I sat in the moonlight on the crest of a dune. The nighttime coolness of the sand seeped through my jeans. Far below me was a round, white, flat bed of silt, spread out on the floor of the Eureka Valley like spilled milk. Around me rose the Eureka Dunes, sickly pale in the moonlight like skin seen underwater. The shadowed hollows between the dunes seemed blacker than the sky above and infinitely deep. The moon was surprisingly bright, so bright I could see the bands of color that rippled across the face of the Last Chance Mountains just east of the dune field: rust, tan, cream, gray—rock colors, desert colors. On the western horizon the Inyo Mountains were black against the sky except where pockets of sand gleamed in the moonlight. A bat flickered in the sky for a moment, then was gone. From the foot of the dune to the farthest visible star, there was no sound.

The bed of silt I saw below me is all that remains of a lake that once spread across the valley floor. The lake bed, now a dry lake—or playa—is probably hundreds of feet thick, built

The steep lee side of a dune at the Eureka Dunes

up over thousands of years from silt and sand eroded out of the surrounding mountains. Once in a while even now, heavy rainstorms in the mountains send water and soil cascading down the boulder-clogged canyons to the lowest part of the valley.

Twenty thousand years ago, the sand that is now piled into dunes lay at the bottom of that lake. About 8,000 years ago, as the climate became warmer, the lake began to dry up. As dry sediments were exposed around the margins of the lake, wind storms picked up the sand grains and blew them about. In the winter, when the wind blew from the east, sand was flung onto the Inyo Mountains on the western side of the Eureka Valley. In the spring and summer, when the wind blew from the west, sand piled up at the base of the Last Chance Mountains, first in small dunes, then in larger and larger ones as small dunes coalesced and wind added still more sand.

Long-leaf sunflower nearly buried by sand

From where I sat, I could not see the highest peak of the dune field, which looms like a tidal wave above the lowest dunes, nor could I see the sandy flat that skirts the eastern edge of the dune field like a beach. But I could see all around me an ocean turned to stone, waves transfixed at the point of breaking, a petrified sea in a landlocked desert.

A light breeze danced across the sand, ruffling my hair. It picked up speed, and grains of sand pattered against my jeans. Then, as suddenly as it had started, the wind stopped. It was enough to remind me, though, that even on the stillest of nights, the essence of sand dunes, like that of the sea, is motion, and that the wind controls the rhythm of sand dune waves as surely as those of the sea.

Just as the ocean's waves can swamp small boats, so sand dune waves can swamp small plants. Even when a dune field is frozen in place by a ring of stable dunes, individual dunes in the middle of the field can still be very active. On the Kelso Dunes,

Arched stems of dune buckwheat show where a dune used to be.

for example, dune crests travel back and forth within a zone thirty to forty feet wide. Within this zone, the total distance traveled by one dune over a period of ten years was several hundred feet, but it ended up only a few feet from where it was first observed. Although hundreds of feet of sand may be excavated by storm winds from a single dune, opposing winds eventually replace the sand so that over the long run, the net change in surface level is measured in inches rather than feet.

Shifting sand is the single greatest challenge plants must surmount if they are to survive on active dunes. During the first three hours of a wind storm, as much as three and one-half inches of sand can accumulate on the windward side of a dune. As much as thirty inches of sand can build up in one place during a twenty-hour gale. It is easy to see why active dunes are often barren. Thirty inches of sand could easily cover most grasses and small shrubs, and many plants would collapse if thirty inches of soil were excavated from their roots. Plants that cannot adapt to moving sand do not persist on active dunes.

Creosote bush, admirably suited for desert life in many ways, is a notable failure on active dunes. On the Death Valley Dunes, creosote grows to be a healthy, vigorous shrub four to five feet tall on both active and stable dunes, but persists only on the latter. Here and there on the lee slopes of active dunes at Death Valley we find pads of sand and flowers and leaves all mixed together. Upon close inspection, the plant fragments turn out to be the tip-top branches of a creosote bush that has been slowly buried alive by drifting sand. After the dune moves on, the knobby, white branches of the skeleton will remain, a tangible reminder of what failure to adapt to moving sand means.

But what about plants that have succeeded? How can they survive where even one of the toughest desert dwellers cannot?

Dune buckwheat, a small shrub that grows on active dunes in the Sonoran Desert, is typical of many dune plants that have adapted to sand movement through rapid growth. Its seeds germinate on dunes when moisture and temperature are favorable. As sand accumulates around the young plant, the main stem elongates rapidly, keeping leaves and flowers above the advancing dune. The stem grows longer and longer as sand piles higher and higher. When the sand is deepest, only the topmost branches appear on the surface, while the elongated stem and root may extend ten feet or more into the sand. When the crest of the dune moves on, the stem, no longer supported by sand, begins to droop. Finally, overbalanced by the crown of leafy growth at the tip, it topples to the ground, forming an arch three, four, or even five feet high.

Dune grass grows only on the Eureka Dunes, and like dune buckwheat and many other dune plants, it too escapes burial through rapid growth. The leafy shoots and rhizomes (underground stems) of dune grass grow especially rapidly in the spring and summer when sand storms are more frequent and intense. Its stems elongate at the phenomenal rate of two and one-half inches a day, keeping leaves and flowers above the

Dune grass builds small dunes of its own on larger dunes.

accumulating sand. As blowing sand is caught and held by its leaves and long, wiry stems, the plant builds a hummock covered by a tangled mat of dead stems, new growth, and exposed roots. Some of these hummocks are up to five feet tall and equally as wide. The hummocks are so tightly bound together by roots and rhizomes that even the strongest winds can seldom tear them apart.

How is such rapid growth possible? In the case of dune grass—and of many other dune plants, as well—high rates of growth go with high rates of photosynthesis. Photosynthesis, of course, is the chemical process plants use to manufacture sugars for growth. Some plants are able to make sugars at a faster rate than others and thus grow more rapidly. Obviously, such rapidly growing plants will be better adapted to active sand than those that grow more slowly.

Many dune plants are unusually large and vigorous, a trait botanists call gigantism. Creosote bush, for example, is a shrub of moderate size in the Mojave, Sonoran, and Chihuahuan deserts, typically growing three to four feet tall. On the sandy edge of the Algodones Dunes, however, mature creosote plants attain heights of six feet or more. Whereas normal fourwing saltbush plants are three to four feet tall and wide, a special form of fourwing saltbush on the Little Sahara Dunes grows ten to twelve feet tall and twelve to fifteen feet wide. Similar examples abound: Thurber's stick-leaf, thorny dalea, squawbush, and desert saltbush all grow larger on dunes. Gigantism in dune plants has several causes: sometimes it is hereditary, as in the fourwing saltbush on the Little Sahara Dunes; sometimes it is due to high rates of photosynthesis, as in dune grass and certain other plants; but most often it is due to the abundant supply of water stored in deeper layers of dune sand.

Cacti are so well suited to the desert in general, it is surprising to find that they are uncommon on all but the most stable of desert dunes. In a way, this is due to the very adaptations that help them survive in the desert. Cacti use a specialized mode of photosynthesis that enables them to conserve water and manufacture sugars at the same time, unlike most plants, which perforce lose water whenever they make sugars. However, cacti and other succulents pay a price for saving water: they grow very slowly. Such slow growth leaves them vulnerable to burial by drifting sand on active dunes, so cacti are rare where sand is mobile.

Rapid growth is crucial to survival of plants on active dunes. Plants that cannot grow quickly enough to outstrip the sand that builds up around them will eventually be buried and killed by moving sand. Yet rapid growth is not the only criterion for success on active dunes.

Plants on active dunes must also be able to form adventitious roots to survive. Just as a cutting of ivy in a glass of water forms roots along the stem, so perennial plants growing on active dunes must form roots along their stems as sand builds up around them.

One of the many dune plants that forms adventitious roots is soaptree yucca, a common plant of plains in the northern

Soaptree yucca on dunes at White Sands

Chihuahuan Desert and on some Chihuahuan Desert dune fields, such as White Sands and the Samalayuca Dunes. In appearance, the soaptree yuccas at White Sands are normal yucca plants. Spheres of lancelike leaves top their shaggy trunks, which hold aloft candelabras of weathered, gray seed pods. But if we were to cut away the sand around the plant, slicing down through the dune to make a gigantic cross-section, we would see that the trunks of the yuccas are not three or four feet tall as they seem, but thirty or forty feet tall—as tall, in fact, as the dune itself. Seeds of the soaptree yucca germinate in swales between the dunes, and because they can form adventitious roots along their trunks, young plants continue growing as they are buried by accumulating sand.

Adventitious roots of soaptree yuccas and other dune plants can reach incredible lengths, often thirty feet or more. Wind-exposed roots, like loose strands of baling wire, are a common sight on active dunes. The root systems in such plants

Wind excavates the roots and stems of soaptree yucca.

are not dense and intricately branched as in nondune plants; rather, secondary roots are widely spaced on the main taproot and have little or no tertiary branching. In addition to transporting water and nutrients from the sand into the plants, such roots actually function like underground vines, providing anchorage in a mobile substrate.

The ability to form adventitious roots is crucial to the survival not only of the soaptree yucca, but of many other plants on active dunes. Failure to form adventitious roots often spells doom for trees that start to grow on active dunes. One of the main problems that trees have when sand builds up around them is oxygen deprivation—literally suffocation. Normally, roots lying just below the surface of the ground supply oxygen to deeper roots. But when accumulating sand changes the ground level, deeper roots are starved for oxygen unless the

Moving sand eventually overwhelms and kills ponderosa pines on the Coral Pink Dunes.

tree can form adventitious roots just below the new surface.

On the Coral Pink Dunes, groves of ponderosa pine trees are scattered across the dunes. In a forest, where trees are crowded and starved for light, ponderosa pines have tall, bare trunks topped by an open crown of branches. But on the Coral Pink Dunes, where light is abundant, ponderosa pines branch near the ground, and their crowns are full, like choice Christmas trees. A casual observer might assume that ponderosa pines are thriving on the Coral Pink Dunes; however, clusters of blackened and leafless tree trunks, like snags after a forest fire, hint at something else.

The Coral Pink Dunes lie at 6,000 feet above sea level, and winters often bring frost and snow. Since wet or frozen sand does not blow about, the dunes advance slowly, only a foot or two each year. They move so slowly that ponderosa pine seeds can germinate in flats between the dunes and grow into trees twenty-five or thirty feet tall before wind-blown sand begins to pile up around the trunks. But given enough time, the dunes inexorably overtake the pines. Unable to form adventitious roots near the sand surface, the pines succumb to slow suffocation. When the crest of the dune has passed, only dead trunks, stripped of branches and needles, remain.

Desert willow, which thrives in broad, sandy washes and floodplains, is one of the few trees that can survive on sand dunes in the arid Southwest. Because the wood of desert willow is soft, the trees bend rather than break as dunes engulf them; because they can form adventitious roots and leafy shoots when submerged in sand, they can survive numerous burials and resurrections. Two more floodplain trees that grow on dunes are honey mesquite and Rio Grande cottonwood. None of these trees are specially adapted to sand dunes, but all can form adventitious roots, an adaptation for survival on floodplains, where occasional floods deposit silt and sand around the trunks. Because they are able to form adventitious roots, desert willow, honey mesquite, and Rio Grande cottonwood are preadapted, as it were, to sand dunes and can thrive in a habitat few other trees can safely occupy.

Wind on sand dunes works against plants in two different ways. Not only does it pile sand around them, it also digs sand from under them. Rapid growth and adventitious roots enable many dune-adapted plants to escape burial by sand but do little to prevent excavation by wind. It is easy to find partially excavated plants on any dune field in the Southwest: a creosote bush that stands on its roots as though propped up on stilts, a palo verde tree with roots radiating out from the trunk like the spokes of an umbrella, a soaptree yucca with a long, naked root like the neck of a giraffe. As wind pries at the roots of such plants, more and more sand is loosened, and eventually the plant loses important feeder roots and dies, or else topples to the ground for lack of anchorage.

Cacti seldom thrive on active dunes, partly because they grow too slowly to outstrip accumulating sand, and partly

Desert willow, a deciduous tree, is one of the few trees that survives on desert dunes.

because their root systems are not well adapted to moving sand. Most cacti have adapted to infrequent rainfall by evolving shallow root systems that quickly absorb even small amounts of moisture as it penetrates the soil. The giant saguaro cactus, for example, which grows to forty feet tall and weighs several tons, has roots that radiate up to one hundred feet from the trunk but lie only inches below the soil surface. Such shallow root systems, although important for survival over much of the desert, make cacti vulnerable to excavation by wind on active sand.

Many sand dune plants, including dune grass, long-leaf sunflower, California croton, and thorny dalea, prevent excavation by building hummocks. Such plants have dense root systems that catch and hold blowing sand, building hummocks through which the plants grow.

Some plants on the dunes at White Sands and at Cuatro Cienegas carry hummock building to extremes, creating

Plants on the Cuatro Cienegas Dunes build hardened pedestals of gypsum sand.

pedestals of sand that are ten to fifteen feet high. Such plants germinate in moist flats between the dunes. As dunes advance, the young plants grow through the sand, binding the gypsum grains with a dense network of underground trunks, branches, and adventitious roots. Eventually, wind disperses loose sand from around the plants, exposing pedestals of cemented sand topped by living plants. Some pedestals are taller than they are wide, others are roughly cubic in shape. Interwoven roots dangle from their walls, which are hard and coarsely etched by wind-blown sand. Squawbush, Rio Grande cottonwood, rosemary mint, soaptree yucca, Mormon tea, and fourwing saltbush are some of the plants that build pedestals of gypsum and thus prevent excavation. After the plants die, the cemented pedestals remain for many years until wind and rain finally erode them into loose grains of gypsum once again.

Moving sand is not the only challenge sand dune plants

face. Lack of nutrients in dune sand is another serious problem. Rotted leaves and other plant parts are an important source of nitrogen in the soil. In the Southwest, desert soils tend to be low in nitrogen because the sparse vegetation produces little organic matter. Although the level of nitrogen in desert soil is low, it is sufficient to support shrubs such as creosote bush and white bursage. Dune sand, however, contains even less nitrogen than typical desert soil. In fact, the small amount of nitrogen in dunes is far below that thought necessary for normal plant growth. Nitrogen is not the only nutrient in short supply in dune sand. The lack of phosphorus is a severe problem for many plants on gypsum dunes, and somewhat of a problem on quartz sand dunes.

Dune plants cope with lack of nutrients in some of the same ways that nondune plants do. Hummock-forming dune plants may concentrate nutrients—particularly nitrogen—in the hummocks. After leaves have fallen to the ground, they are often trapped in the hummock where they decay, releasing nitrogen and other nutrients into the root zone. Some plants have a symbiotic relationship—one that is mutually beneficial—with certain bacteria: the bacteria obtain energy from the host plant and in return supply nitrogen. Such bacteria are called nitrogen fixers. They are able to transform atmospheric nitrogen, which forms seventy-nine percent of the air we breathe, into forms of nitrogen that can be used by plants. Living in nodules on the roots of the host plant, these bacteria can supply a substantial proportion of the nitrogen required by the host. The root nodules are visible to the naked eye, but the bacteria themselves are microscopic in size.

Nitrogen fixation is characteristic of the legume family, and among the dune plants that fix nitrogen are scurf-pea, silvery sophora, locoweed, and Fremont dalea, all legumes. Several dune grasses also use nitrogen-fixing bacteria. Instead of forming root nodules, these grasses—Indian rice grass, thickspike wheatgrass, needle-and-thread grass, and purple three-awn—develop sandy sheaths around the roots. The sheath is formed as the root tips exude a slimy glue that cements grains of sand to the root. Nitrogen-fixing bacteria live inside the sheaths, where soil moisture and organic matter are high compared with the sand outside.

Another problem dune plants face is high temperatures. When leaf temperatures are too high, photosynthesis shuts down. Heat on sand dunes comes from several sources: visible light, that is, those wavelengths of sunlight that the human eye can see; infrared light, which is not visible to the human eye; and reflected light, as waves of visible and infrared light bounce back from the surface of the sand.

Many species of dune plants have adapted to excess light and heat by evolving silvery, white, or gray leaves. In some species—dune sunflower, for example—a dense coat of fine hairs gives the leaves a silvery sheen and a velvety touch. In others, such as Wiggins' croton, overlapping scales like tiny plates cover the leaves. Hairs or scales help plants regulate the temperature of their leaves by blocking out some of the light, and thus the heat, in their environment. Leaves of hoary dicoria, a hairy gray annual found on many southwestern dune fields, absorb only sixty-six percent of the light that strikes them. In contrast, the green and hairless leaves of creosote bush absorb nearly twenty percent more light than those of hoary dicoria. Leaf hairs or scales not only help keep leaf temperatures low enough for photosynthesis, they also help to slow evaporation from the surface of the leaf, thus retarding water loss.

Ecologists have found that the hairiness of some plants varies during the year. After spring rains the leaves are nearly green, but as the soil dries out they become hairier and more silvery, and thus more reflective of light. Their level of photosynthesis corresponds with hairiness: when leaves are green, photosynthesis is high, and when leaves are gray, photosynthesis is low. Palmer's brittlebush absorbs as little as forty-five percent or as much as eighty-two percent of the light that strikes its leaves, depending on how dense the coat of hairs is. By adjusting the amount of light that penetrates the leaves, plants can maintain a delicate balance with their environment,

Dune sunflower

producing sugars for growth no matter what the air temperature is, since the reflective leaves keep leaf temperatures at the optimum level for photosynthesis. Such a balance is particularly important for dune plants, since sand is highly reflective. Among the many dune plants with conspicuously hairy leaves are silvery sunflower, dune sunflower, Wiggins' croton, California croton, silvery locoweed, Peirson's locoweed, dune peabush, dune buckwheat, hoary dicoria, rosemary mint, sand sage, and sand food.

In summary, dune plants have evolved an impressive array of adaptations to their environment. Adaptations to moving sand are crucial to the survival of plants on active dunes. Most plants on active dunes grow rapidly, particularly as seedlings. As dune plants are buried by sand, their stems elongate through the sand, while adventitious roots sprout along the buried stem. The adventitious roots of some dune plants stretch for yards through the sand, both anchoring the plant and seeking water and nutrients for it. In other dune plants, roots and underground stems form dense hummocks, preventing excavation of the roots by wind. Some dune plants—mostly members of the pea family—supply their own nitrogen by fixing atmospheric nitrogen in the sand. To avoid overheating caused by sunlight and reflection, many plants don a silvery, white, or gray coat of hairs or scales. The hairs block out some of the excess sunlight, enabling the plant to maintain leaf temperatures at optimum levels for photosynthesis.

In spite of the many ways sand dune plants cope with their environment, survival is not assured. Seedlings are particularly vulnerable to the rigors of life on dunes.

Double Bind

WHEN APPROACHED FROM THE EAST, THE ALGO-dones Dunes seem to stretch across the entire horizon, peak after peak of sunburnt sand rising hundreds of feet above the desert floor, an endless chain of pyramids shimmering against the electric-blue sky. These pyramids were not built from blocks of stone, of course, but from grains of sand. According to geologists, the Algodones Dunes are complex dunes: each sand pyramid is itself a dune surmounted by others, dune upon dune leading like stairsteps from one landing to the next.

The sand source for the Algodones Dunes was most likely sediments deposited in Lake Cahuilla, a much larger fore-runner of the Salton Sea. Sand at the source must have been plentiful, for the dune field is forty miles long and four to six miles wide.

My first visit to the Algodones Dunes coincided with a bumper crop of seedlings. Although it was early spring, the days had been warm and nights had been mild. A late winter storm, blowing east across the desert from the Pacific Ocean, had dropped over half an inch of rain on the dunes about two

The Algodones Dunes

weeks prior to my visit. Although the sand was warm and dry on the surface, it was cool and moist just a few inches underneath. Stimulated by warm temperatures and timely rainfall, many seeds had recently germinated, and seedlings were scattered over the dunes, sometimes as many as fifteen or twenty in a spot the size of a quarter. I found seedlings of every species of shrub that grows in the central part of the dune field, thousands of seedlings in all.

Later on, walking across the desert plain east of the dune field, I could not help but be struck by the contrast between the dunes and the flats. Although seedlings were plentiful on the dunes, I found hardly any on the flats—no seedlings of ironwood or little-leaf palo verde or ocotillo, even though these are common plants on the gravelly flats.

Most desert perennials do not reproduce abundantly. Even when a good crop of seeds is set, it might be many years until they germinate, and when they finally do, few survive past the first year. The age distribution of perennial plants on the desert plain—many more mature plants than seedlings or juveniles—is typical of arid environments. Full-grown plants intercept and use what little water there is, leaving none for seedlings, except in years of abundant rainfall when there is enough to go around for a little while.

The age distribution of plants on the Algodones Dunes, where the number of seedlings often exceeds the number of mature plants, is more typical of environments where water is abundant. Since dune sand is moist under the surface, almost any spot where seeds of dune-adapted plants land is favorable for growth, given the proper temperatures and enough rainfall for germination. Even though it seldom rains in the desert, most of the rain that falls on sand dunes is stored until a plant uses it or until it percolates into the water table. The difference between the age distribution of perennials on and off the dunes is due largely to soil moisture; the abundance of seedlings and juvenile plants on the dunes shows that water is more available there than on nearby gravelly flats.

With so many seedlings on active dunes, it seems likely that plants would eventually cover the dunes, yet they do not. A spur-of-the moment experiment made at the Algodones Dunes helped explain this paradox. I pounded stakes into the sand at several different locations and made a small map of the seedlings around each stake. When I returned two months later, I found that a number of things had happened. First of all, and not surprisingly, the sand had shifted: as much as six inches of sand had accumulated at some of the stakes. In addition, the top six inches of the sand had dried out. Soil moisture had dropped from an average of 0.36 percent to 0.19 percent. But most important of all, more than half the seedlings had not survived. Presumably they had been buried by shifting sand. This crop of seedlings was caught in a double bind; accumulating sand made it necessary for them to grow, but lack of water in the root zone made growth impossible. Only rain could have stayed their inevitable burial, but no rain came.

Giant Spanish needles

The decimation of the seedling population in just two months not only shows that sand dunes are tough environments for seedlings, but also demonstrates that most seedlings on the dunes never reach maturity. That may be one reason why desert dunes are not covered with plants: most seedlings never make it past their first year.

Still, those plants that do reach maturity live in one of the least desertlike of desert habitats, not just on the Algodones Dunes, but on all desert dunes. It cuts both ways, though. Just as widely distributed desert plants such as creosote bush and white bursage cannot survive on active dunes, plants adapted to moving sand cannot survive anywhere else. Their root systems, although admirably suited for anchorage in loose, sandy soils, lack enough surface area to compete for water in clay soils with densely rooted nondune plants. Plants such as dune sunflower and giant Spanish needles have, in effect, made a choice involving a substantial trade-off. By adapting so completely to moving sand, they live essentially free of competition with other plants. For this freedom, they have sacrificed the opportunity to grow in other habitats. They have traded a broad but competitive niche for a secure but small one.

We have seen that mature dune plants grow rapidly through accumulating sand, forming adventitious roots as their stems are buried. We have also seen that the most dangerous period in a dune plant's life is the seedling stage, when its roots might not be able to reach past the dry layer of sand near the surface to moist sand underneath. But this is only part of the story. To survive on dunes, plants must adapt to moving sand at every stage of their lives, not only as mature plants and seedlings, but also as seeds.

Little Oaks from Great Acorns

THE SAMALAYUCA DUNES, LIKE THE KELSO AND Algodones dunes, are a mountain range made of sand. Six major peaks, strung together like beads on a chain, rise more than five hundred feet above the valley floor. Many square miles of low, rolling dunes and sandy plains surround the larger peaks. Behind the dune field to the northeast lies the Sierra del Presidio, a long, low, desert mountain range whose crest is jagged, as though carelessly torn from paper. At midday, the range seems flat and fake, a piece of stage scenery propped against a blue-painted backdrop. To the southwest lies another mountain range, the Sierra de Samalayuca. It is in the broad valley between these two ranges that the tallest dunes lie.

The ninety-seven square miles of sand that make up the Samalayuca dune field are part of a much larger intermittent sand sea that stretches from northern Chihuahua to western Texas and southeastern New Mexico. Geologists think that several million years ago, a vast lake, Lake Cabeza de Vaca, covered much of this region. The lake, which was fed by the ancestral Rio Grande, lacked an outlet to the sea, and like some mythical monster, passively swallowed everything that was fed to it. Mountain ranges became islands in the lake as the water level rose. At its greatest extent, the lake may have covered 9,000 to 10,000 square miles. Finally about 800,000 years ago, the lake level rose so high that it breached the Quitman Mountains east of El Paso, and the waters drained away, carried by the Rio Grande to the Gulf of Mexico. Instead of a single immense lake, as before, there remained a series of smaller lakes, each in its own basin between the mountains.

As the climate became warmer and drier all over the Southwest, these smaller lakes evaporated, leaving behind playas and thick accumulations of sediment that had been washed down from the mountains of New Mexico and Colorado by the Rio Grande. Sand from the playas was blown into the Samalayuca Dunes and into the sand hills near El Paso and Las Cruces.

The Samalayuca Dunes are at their best in August and September after summer rains have soaked into the wildflower seeds that lie dormant in the sand. Then the margins of the dune field burst into bloom: spreading mats of pale evening primrose, blowsy bushes of bindweed heliotrope, upright stalks of palafoxia. By late fall no trace of summer's flowers remains: the blossoms have faded and fallen, the stems have withered and blown away. Just below the surface of the sand, though, amid the myriad shifting grains, a new crop of seeds waits for the rains of the coming summer, or if no rains come, for those of the summer after that.

Ecologists often talk about reproductive strategies, that is, the way a plant or animal ensures that reproduction will succeed. Annual wildflowers on the Samalayuca Dunes show one type of reproductive strategy. They are ephemerals, and their strategy is to germinate, flower, and bear seeds in a single, favorable season, to grow in the wake of summer rains and exist as seeds at other times of the year when the climate is too cold or too dry.

One of the most important components of an ephemeral's

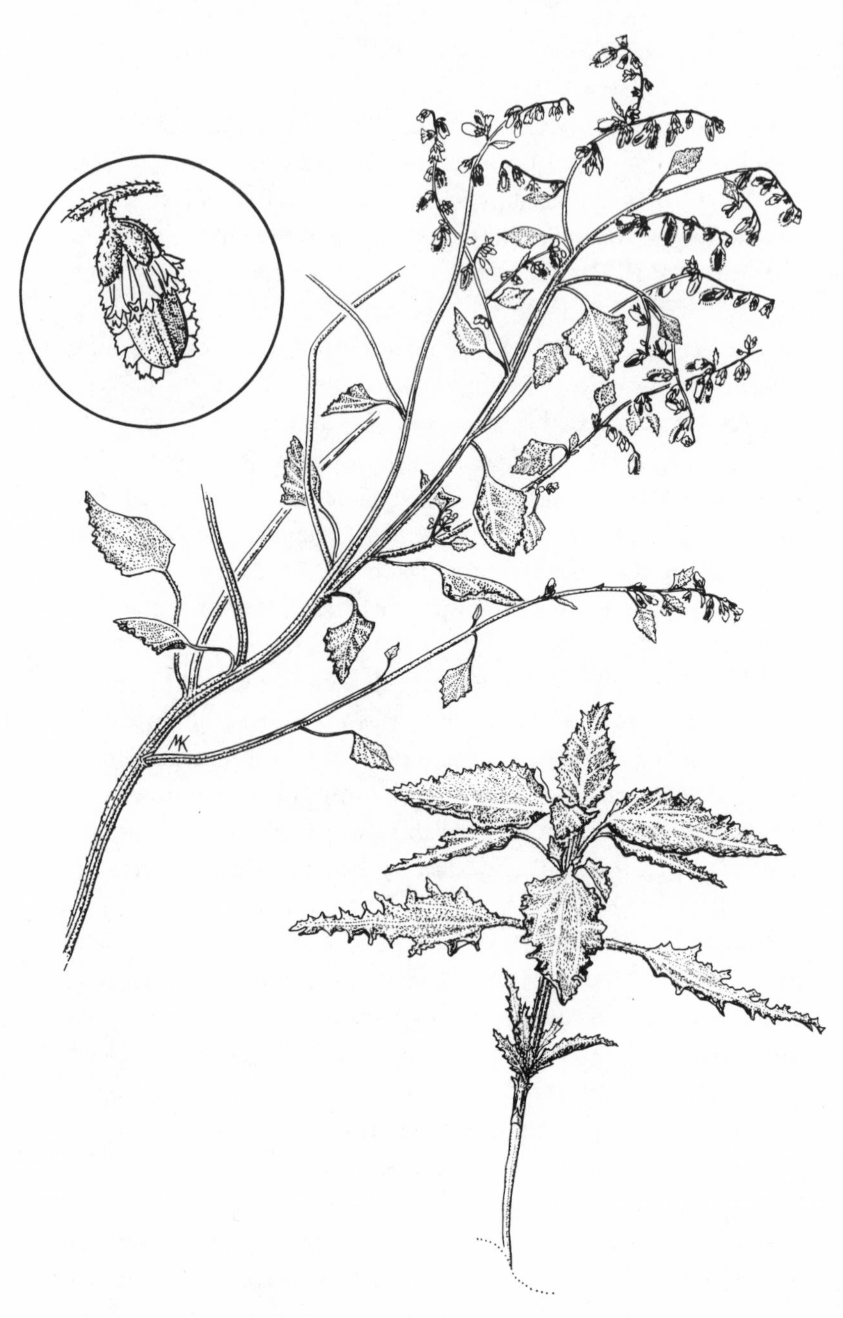

Hoary dicoria

strategy is prolific seed production, as shown by hoary dicoria, a robust annual that grows on dunes throughout the Sonoran and Mojave deserts. Seeds of hoary dicoria germinate in the spring after winter rains. Since spring is the windy season in the Mojave and Sonoran deserts, seedlings must grow rapidly. By midsummer, the plants are often three feet tall with roots reaching over four feet into the sand. (Like dune grass, hoary dicoria has a high rate of photosynthesis, which sustains rapid growth.) The plants go to seed in the summer. Often they remain anchored in the sand during the fall and winter while winds disperse the papery bladders that enclose the seeds; but sometimes the stems break off above the sand and the plants tumble across the dunes like beach balls, scattering seeds as they go. The following spring, the seeds will germinate wherever they land: in swales, on windward slopes, even on the steepest lee faces of the dunes.

Successful reproduction is not guaranteed, however. The seeds could be blown off the dunes altogether, since the bladders fly so readily in the wind. They could be buried too deeply to emerge before they have exhausted their meager supplies of stored food. Many of the seeds are eaten by rodents. Hoary dicoria overcomes these hazards to some extent by producing large numbers of seeds. As long as seed production outweighs seed destruction, some seeds will be left to carry on the species.

A different strategy is followed by scurf-pea, a perennial herb that grows on the Coral Pink Dunes and other dune fields in the Great Basin Desert. Scurf-pea not only bears fruits and reproduces from seed, it also reproduces vegetatively. Like many plants on active dunes, scurf-pea grows from rhizomes. As wind-blown sand covers these whiplike underground stems, adventitious roots and leafy shoots develop at one- or two-foot intervals, creating new plants.

Reproduction by seed involves recombination of genetic material from each parent plant. Vegetative reproduction involves no genetic recombination; the new plant is a clone of the parent. Each strategy has advantages and disadvantages as far as dune plants are concerned.

The biggest advantage to reproduction by seed is that offspring produced by genetic recombination differ from one another in unpredictable ways. Such differences could help the species survive if the environment changed. Although most offspring would not survive severe upheavals, a few might be adapted to the new conditions, and those few could make the difference between survival and extinction for the species. Plants that reproduce vegetatively lack such genetic variation. They could perish if their environment changed drastically.

On the other hand, reproduction by seed is hazardous in any environment and especially so on dunes. Seeds could be buried too deeply by moving sand, blown off the dunes altogether, or eaten by ants, birds, or rodents. To germinate, seeds of desert plants must often wait for just the right combination of temperature and rainfall.

Plants that can reproduce vegetatively are not limited by such constraints. For one thing, vegetative reproduction ensures that the plant will stay where it belongs, that is, on sand. For another, blowing sand, in covering rhizomes and rootstocks, enhances the formation of adventitious roots and shoots and thus of new plants, while seeds lie dormant in the soil. In addition, when the environment is stable, vegetative reproduction preserves the genotypes that are already well adapted to the prevailing conditions.

Many perennial dune plants, such as dune evening primrose, dune grass, and scurf-pea, reproduce both vegetatively and by seeds. They alternate between strategies according to rainfall, temperature, and sand movement and thus ensure that they will persist.

Because the sand dune habitat is rigorous not just for mature plants, but for seeds, as well, seeds also exhibit adaptations to their environment. One obvious adaptation is that seeds of many dune plants are larger and heavier than the seeds of their nondune relatives. Peirson's locoweed has the largest seeds of any locoweed in the Southwest. Shinnery oak is

Shinnery oak

small—it grows only knee- to waist-high—but its acorns are surprisingly large: an inch long and a half-inch thick. Dune sunflower has longer and heavier seeds than the annual sunflower that grows on stabilized dunes in the Sonoran Desert. The seeds of plains yucca are half again as long as many other yucca seeds.

Bigger seeds in dune plants are an important adaptation to burial by sand. Small seeds will germinate when buried under several inches of sand, but cannot emerge above the surface before exhausting their meager supplies of stored food. Larger seeds store more food, enabling the newly germinated seedling to elongate through even several inches of sand and emerge before the stored energy in the seed is depleted.

Large seed size helps a plant survive once the seed has landed on a dune field. Seeds arrive on dunes by a variety of means collectively called dispersal mechanisms.

The mechanisms plants have for moving their seeds from place to place are endless: wings, barbs, glue, hairs, and more. For the sake of simplicity, ecologists often group these mechanisms according to dispersal agents, including wind, water, animals (anything from an ant to a bird to a cow to a human), the plant itself, no dispersal at all, and dispersal in time (which overlaps some of the other categories). Seeds dispersed by wind—the agent—typically have a parachute of hairs or a flattened wing or two—the mechanism. Some seeds dispersed by animals—another agent—adhere to fur or feathers or socks with barbed hairs or stickers—more mechanisms. Others have a fleshy coating like a plum so that animals will eat them and pass the seeds unharmed through their digestive tracts. Seeds dispersed by water may be buoyant, with air pockets to keep the seed afloat and a thick seed coat to prevent damage to the tender embryo inside. Many plants have capsules or pods that uncurl or pop open so that the seeds scatter a short distance from the parent plant. (Such seeds may be transported still farther if an animal eats them, an unavoidable overlapping of categories.) Some seeds—those with no dispersal mechanisms and no dispersal agents—simply fall close to the parent plant and go nowhere.

Interestingly enough, seeds that go nowhere, or almost nowhere, are typical of deserts, especially Old World deserts. These seeds lack any particular adaptation for long-distance dispersal. They have no tufts of hair that can be caught by the wind, no barbed bristles that can catch on a wing or a hide or a shirtsleeve, no fleshy covering that makes them good to eat.

Two theories account for the lack of long-distance dispersal in desert plants. One theory states that the parent plant, having found a good spot in which to grow and reproduce, can ensure the survival of its offspring by scattering its seed closely around itself. The other theory suggests that seeds of desert plants are not dispersed long distances simply because there is no particular advantage to it. For one thing, the environment in a desert varies more through time than it does through space. Think about winter rainfall in the Sonoran Desert: one year may be wet, whereas the next three or four may be dry. Whether the rains are good or scanty, the moistness or drought is likely to extend over many square miles. It does a seed no good to be dispersed from one drought-stricken place to another, but it *is* useful to the seed to be dispersed from a drought-stricken period to a wet one. A seed that disperses through time will hit favorable conditions sooner or later. Seeds that apparently go nowhere upon dispersal are often dispersing through time.

The mechanism for dispersal in time is seed dormancy. Many desert wildflowers produce seeds that can lay in the soil for years before germinating. Such seeds have hard seed coats that must be abraded by rocks and gravel before they will germinate. Some seeds contain chemicals that prevent germination until a good rain washes the inhibitor away.

Seed dormancy is a way for the species to escape a series of drought years, postponing germination until conditions for survival are more favorable. In any given year, perhaps a few seeds will germinate. But what if these seeds wrongly predicted their chances for survival and die because there is too little rain after germination, or a sudden freeze, or a prolonged heat wave? In a species having seed dormancy, some seeds will still be in the soil, waiting until a better year comes along. This way the parent plant disperses its seeds through time, improving the chances that at least some of its offspring will survive and reproduce.

All of these dispersal types are represented among desert dune plants in the Southwest, although not all are equally represented. Plants whose seeds are adapted to wind dispersal are often widely distributed, and on dunes, they are also likely to be annuals. Hoary dicoria—found on the Gran Desierto, Algodones, Kelso, and other dune fields—is one such plant. The seeds of hoary dicoria are well suited for dispersal from one dune field to another: the leaflike bracts around the ovary become broad, papery wings as the seed matures, and the entire structure is readily floated aloft by the wind. Sand verbena, another widespread annual, grows in sandy soil and on stable dunes in the Mojave and Sonoran deserts. Like hoary dicoria, the seeds of sand verbena are winged, an adaptation for wind dispersal. Dispersal by animals does occur among desert dune plants but not as commonly as other dispersal mechanisms. Squawbush, which grows at White Sands, has berrylike fruits that are relished by birds. Various wolfberries, mistletoes, and cacti have fleshy fruits that are eaten by animals, and thus spread from one stabilized dune to another. Mesquite is often spread by cattle, who eat the pods and excrete the hard-coated seeds in a package of fertilizer. Seeds of some dune plants are collected by ants and rodents and, if cached but not eaten, may germinate and grow to maturity.

Dune annuals that exhibit seed dormancy include evening primrose, sand verbena, spectacle-pod, and Spanish needles. Some perennial dune plants probably also exhibit seed dormancy, but little is known about them. Seeds of some plants growing on coastal dunes are dispersed by ocean currents.

There is much overlap between the categories of self-dispersal and no dispersal. Some dune plants hedge their bets by doing both. Evening primrose, a common annual wildflower on stable dunes and sandy plains in the Mojave and Sonoran deserts, is a classic example. When its capsules pop open, some seeds fall from the top half, while others stay tucked in the bottom. Drifting sand eventually buries the skeleton with its cache of half-filled capsules. The next spring that has good rains finds seeds in the capsule germinating in place, outlining the buried skeleton of the parent plant. Evening primrose hedges its bets by dispersing some seeds a short distance away and keeping others at the exact spot that worked once before.

Self-dispersal or no dispersal at all is the typical condition for most desert dune plants, particularly for perennials that grow on only one or two dune fields. Dune sunflower, for example, has small, wedge-shaped seeds that have no special adaptation for dispersal. As in many sunflower seeds, there are a pair of bristles on the apex of the seed, but these fall off by the time the seed is mature. The seeds of Wiggins' croton are of marblelike smoothness, with no appendages at all. When its capsules split open, the seeds fall several feet, at most, from the parent plant. Giant Spanish needles has seeds that are crowned with a series of eight parchmentlike bristles. Although these should be of some use in parachuting the seeds through the air, the seed itself is long and heavy, making the whole structure less buoyant and less likely to be dispersed very far by the wind.

Seeds that drift off the dunes and land on the inhospitable and sandless desert floor are wasted. The lack of adaptations for long-distance dispersal in these and other dune plants does serve a purpose: it confines them to the deep, active sand to which they are best suited.

Changes

THE DUNES AT WHITE SANDS ARE LONG AND LOW and rounded, like giant whales beached and turned to stone, a graveyard of fossil cetaceans. We arrived there as a summer storm was building over the San Andres Mountains in the west. Black clouds cantered across the sky, trailing a fringe of rain beneath them, while we sweated in the sunshine, nearly blinded by light reflected from the snow-bright surface of the sand.

White Sands lies in the Tularosa Basin in southern New Mexico at an elevation of about 4,000 feet, bounded on the southwest by Lake Lucero, on the west by the San Andres Mountains, and on the east by the Sacramento Mountains. Lake Lucero is the immediate source of sand, which is nearly pure gypsum—a very soft mineral that is white or almost colorless when pure. The lake is a playa, the Spanish word for a lake that is dry except after rainstorms. Sediments in Lake Lucero and its Pleistocene precursor, Lake Otero, eroded out of gypsum rock formations in the Sacramento and San Andres mountains. Twenty to thirty feet of ancient dunes underlie the

Active transverse dunes at White Sands

present dune field, and this ancient dune system lies on silt and gravel washed down from the mountains.

Dunes originate as strong winds blowing across Lake Lucero lift up grains of gypsum and pile them into low, dome-shaped dunes on the west side of the lake. Dome dunes advance rapidly to the northeast, where they merge with active transverse dunes in the center of the dune field. The transverse dunes are low, usually no more than forty feet tall, but they make up for their lack of height in breadth and depth: some are up to eight hundred feet long and four hundred feet across. Parabolic dunes, securely fixed by vegetation, line the eastern and northeastern sides of the dune field. A striking feature of White Sands is the broad flats—called swales—between the dunes. These are most prominent between the parabolic dunes, where the area occupied by the swales actually exceeds that covered by the dunes themselves. Swales are ephemeral, disappearing as dunes creep over them and gradually reappearing as the dunes move on.

At White Sands the different types of dunes support different mixtures of plants, or plant communities. The dome-shaped dunes are quite barren, but a few plants—alkali sacaton, Indian rice grass, pale evening primrose, and narrow-leaved sand verbena—grow at widely spaced intervals between the dunes. The transverse dunes are dotted with squawbush, rosemary mint, soaptree yucca, and Mormon tea. Some of these plants perch on pedestals high above the dunes, others sit on low hummocks. The parabolic dunes are well anchored with plants: Mormon tea, rosemary mint, squawbush, soaptree yucca, fourwing saltbush, crucifixion thorn, Rio Grande cottonwood, rubber rabbitbrush, claret cups hedgehog, and plains prickly pear.

The difference in plants between the barren dunes near Lake Lucero and the vegetated dunes on the east side of the dune field *might* be ascribed to succession, the replacement of one group of plants by another due to the action of plants on their environment. On sand dunes the most important action leading to succession is stabilization of dunes by plants. As the first few plants colonize active dunes, sand movement decreases, enabling a few more plants to gain a foothold. This in turn slows dune movement still more, so that additional plants become established on the sand, and finally the dunes are completely stalled.

Ecologists developed the classic model of succession in moist, humid climates. According to their scheme, bare rock is colonized by lichens, which eat away at the substrate, forming pockets of soil. Mosses and ferns follow, adding to the organic matter in the soil and making it rich enough for grasses and tough, weedy wildflowers. Shrubs are the next occupants, and in their shade, tree seedlings find enough protection from the sun to sprout and grow. Sun-loving trees are followed by shade-loving trees, and eventually a forest grows where there was once only rocky ground. This is primary succession, that is, succession starting where no plants grew before. In reality, we seldom see primary succession in nature. More often, we observe abandoned farm fields becoming weed and wildflower patches, then woodlots, and finally tracts of forest. This is secondary succession, that is, succession starting after some disturbance to the original plant community.

The classic scheme of succession is not always a useful concept for interpreting patterns of vegetation in deserts. Although many plant ecologists have said that succession does not happen in deserts, scientists working in the northern Mojave Desert have recently found secondary plant succession on the sites of old ghost towns. After settlers abandoned these towns, weeds such as Russian thistle, red brome, and skeleton-weed colonized the soil where roads and tent buildings once stood. Over a period of fifty years or so, Mormon tea, cheese-bush, needle-and-thread grass, and Indian rice grass replaced the weeds. Eventually, perhaps after one hundred years or more, lightly used roads will be nearly indistinguishable from the nearby desert and will support creosote bush, blackbrush, and hopsage. Heavily used roads might not return to normal for a thousand years or may never completely recover. Does the White Sands dune field show this kind of pattern? Perhaps not.

Successional patterns on coastal dunes in Oregon help to clarify the issues at hand. Along the Oregon coast, stable dunes support forests of lodgepole pine and Sitka spruce, with a shrubby understory of huckleberry, rhododendron, kinnikinnik, bracken fern, and other shrubs and ferns. Occasionally, advancing sand overwhelms the forest community, killing virtually every plant in it, since they are not adapted to moving sand. Secondary succession follows, as perennial herbs—yellow sand verbena, beach strawberry, coastal bursage, and marram grass—colonize bare sand. After herbs have slowed sand movement, certain shrubs and even a few ferns can germinate and thrive: these include shallon, kinnikinnik, huckleberry, and bracken fern. Lodgepole pine seeds germinate well in bright light and mineral sand, so a few pines also begin to grow on the dunes. As the pines grow taller and the shrubs become denser, more and more seeds of the shade-loving Sitka spruce germinate and grow to maturity. Gradually, over a period of one or two hundred years, the once-barren dune reverts to the original lodgepole pine and Sitka spruce forest.

Another vegetational pattern occurs on these same coastal dunes. Loose sand and wind build small hummocks—called foredunes—next to the ocean. On or behind the foredunes grow a handful of plants that can tolerate not only moving sand but salt spray and desiccating coastal winds. These plants include yellow sand verbena, beach morning glory, beach silvertop, big-head sedge, and coastal lupine. But this little community does not prepare the way for huckleberries, rhododendrons, and lodgepole pines because these three species cannot tolerate salty ocean spray. The only way a pine and spruce forest could grow where the foredunes are would be if the tideline receded enough to give the forest trees and shrubs protection from salt spray and wind. Obviously, this would involve major changes in the landscape, such as a lower sea level or an uplift of the land.

Dunes along the Oregon coast (and along much of the California and Washington coasts) thus show two kinds of vegetational patterns. The first pattern is the classic succession through time. The second pattern is zonation, not succession; there is a zone along the immediate coast that will not support a pine and spruce forest in our lifetimes or our children's lifetimes or their children's lifetimes. In fact, if a pine and spruce forest is to succeed coastal foredune herbs, immense spans of time—geologic time—are necessary. This is not succession as it is commonly defined or understood.

How do these two patterns relate to desert dune fields? We can most easily detect the classic scheme on dune fields where sand is no longer supplied to the system. For example, stable dunes east of the Colorado River near Bouse, Arizona, were probably built from river sand thousands of years ago. No more sand is being added to these dunes, which are well anchored by white bursage, big galleta grass, and Mormon tea, yet at one time these must have been active dunes, colonized by a handful of species able to tolerate moving sand. Similarly, the Algodones Dunes, built from sandy sediments of Lake Cahuilla, are no longer being supplied with a fresh source of sand. Creosote bush, Mormon tea, dune peabush, white bursage,

Big galleta grass stabilizes marginal dunes at the Kelso dune field.

and big galleta grass have stabilized the perimeter of the dune field. None of these grow in the central dune field, where sand is most active: here we find fast-growing plants such as dune sunflower, giant Spanish needles, d'Urville's panic grass, Wiggins' croton, dune buckwheat, and hoary dicoria. Given enough time, the center of the dune field will become stable too, just like its margins. Plants adapted to active sand might survive on blowouts, areas where stable sand works loose and engulfs nearby plants. The Algodones Dunes also fit the classic scheme of succession: plants typical of stable sand gradually replace those typical of active sand, and the sparse vegetation becomes denser.

The Mohawk Dunes in the southwestern corner of Arizona are midway between the Algodones and Bouse dunes in their successional development. Like the other two dune fields, the Mohawk Dunes have garnered all the sand available for their formation; no more sand is being blown into the dune field. Although creosote, Mormon tea, big galleta grass, and white bursage dominate much of the dune field, blowouts support a few species typical of active sand, including hoary dicoria and dune buckwheat.

In contrast, the dunes at White Sands are still being supplied with sand from Lake Lucero, and the dune field consists of active, partly stable, and completely stable dunes. The vegetation reflects this pattern, as we have seen: dome-shaped dunes next to Lake Lucero are barren; active transverse dunes some distance away from the lake support scattered plants; and stable parabolic dunes at the far edge of the dune field support a variety of shrubs, herbs, grasses, and cacti. Perhaps this pattern is more analogous to the zonation on foredunes along the coast than to classic plant succession. Just as forest trees cannot tolerate the salinity of coastal foredunes, so the plants on stable dunes at White Sands cannot tolerate the active sand next to Lake Lucero. As long as more and more sand is blown into the dune field, dome-shaped dunes will remain barren, active dunes unstable. The dunes at White Sands show a pattern of zonation in space rather than succession through time.

Of course, none of these patterns are immutable. A long series of drought years could eliminate many plants on stable dune fields, freeing the sand to blow about and reactivating the dunes. A long series of wet years would probably shift the balance the other way, hastening stabilization. Cycles of wet years and dry, of moving dunes and anchored ones, are the essence of dune country. Change, after all, is what sand dunes are all about.

Nowhere Else in the World

CUATRO CIENEGAS—FOUR MARSHES, LOOSELY translated—is a horseshoe-shaped basin in the center of the Chihuahuan Desert. At the north end of the basin on the neck of the horseshoe, surrounded by corn fields and stockyards, is the town of Cuatro Cienegas, a dusty, desert town of 14,000 inhabitants. Miles and miles to the south, out of view from the town, across rush-filled marshes and grassy plains where horses and burros graze, past stony bajadas ornamented with cacti and thorny shrubs, over a flat, blue lake that stares unblinking at the sky above, lies the Cuatro Cienegas dune field, eight square miles of white gypsum sand. The dunes lie west of the lake, Laguna del Churince. Like the other small lakes in the basin, this one is fed by a little stream that originates in a spring-fed pond.

When I first saw the Cuatro Cienegas Dunes, I was struck by their similarity to White Sands: the same bright, white gypsum sand, the same gypsum pedestals like the ruins of ancient temples, the same flat-topped, steep-sided parabolic dunes. As at White Sands, the sediments in Laguna del Churince eroded out of gypsum-laden formations in nearby mountains. As at White Sands, wind blew dried lake sediments into the dunes that lie west of the lake. While I wandered across the dune field, however, I began to notice more differences than similarities between White Sands and Cuatro Cienegas.

At White Sands, the dune plants were like familiar friends. I recognized many of them from other places in the Southwest: Rio Grande cottonwood, rosemary mint, Mormon tea, soaptree yucca, and so forth. But at Cuatro Cienegas, the plants were strangers. I recognized some—honey mesquite, catclaw, desert willow, fourwing saltbush, and a few others—but most were new to me. I saw yuccas, but they were not soaptree yucca; I saw a hedgehog cactus, but it was not claret cups hedgehog; and there were other plants I did not know at all, plants I had never seen on any other dune field.

Later I learned there was a good reason why many plants on the Cuatro Cienegas Dunes seemed like strangers to me. Many grow only on gypsum soils in the Chihuahuan Desert. Three others grow nowhere else in the world but on the Cuatro Cienegas Dunes. Botanists call such plants endemics, meaning that they are restricted to a relatively small area or to a particular type of soil.

Endemism in the Cuatro Cienegas region is not limited to plants. Nine species of snails, three of turtles, three of lizards, five of scorpions, and eight of fish are also endemic to the basin. Why are so many species restricted to this one area? The question has interested biologists for years.

In part, plant endemism in the Cuatro Cienegas basin is due to the gypsum that makes up the dunes, the playas, and the salt flats on the basin floor. Many plants cannot live on gypsum, for reasons that are not entirely clear. Some biologists think that gypsum contains too much sulphur for most plants. The plants that *can* grow on gypsum are of two kinds: those that merely tolerate gypsum and those that actually prefer it or even require it. Plants that require gypsum are, by definition, gypsum endemics. Biologists assume that gypsum endemics evolved from ancestors that grew near but not on gypsum soils.

In part, endemism in the basin is due to time. Although we find gypsum deposits all the way from Mexico to Montana, those in Mexico are rich in endemic plant species, whereas those in Montana have none. During the last Ice Age, when glaciers covered gypsum deposits in Montana and other northern states, those in Mexico were free of ice. Since the gypsum deposits in Mexico have been exposed and undisturbed for a longer period of time, there has been more time for the evolution of endemics there.

Gypsum soils and long periods of time do not, in themselves, solve the puzzle of endemism in the Cuatro Cienegas basin. The piece that completes the puzzle is the element of isolation. Because high mountain ranges surround the Cuatro Cienegas basin, special habitats such as springs, dunes, and salt marshes are isolated from similar habitats in other basins. Isolation limits gene flow between populations of closely related species, enabling them to maintain the differences that make them distinct species.

But how, exactly, did the three species of plants endemic to the Cuatro Cienegas dune field—contorted aster, gypsum blanketflower, and gypsum stinkweed—evolve, given gypsum, time, and isolation?

Contorted aster is the most bizarre of the three. Most asters are stiff and upright, but contorted aster is prostrate and twisted. Clusters of short flowering stems spring up at intervals from creeping, woody branches that lie half-buried in drifting sand. Tiny, triangular leaves tightly sheathe the stems, and crinkly, woolly hairs cover the leaves and stems.

It is not hard to imagine how contorted aster evolved on the dunes, if you keep in mind that all plant populations are variable, just as human ones are, and that this variation is partly controlled by genes. Picture a population of asters, all perennial herbs, growing near the dunes. Some have stiff, upright stems and green leaves. Others have loose, more flexible stems and hairy leaves. As sand blows over these plants, the ones with stiff stems and green leaves suffer. Their stems break

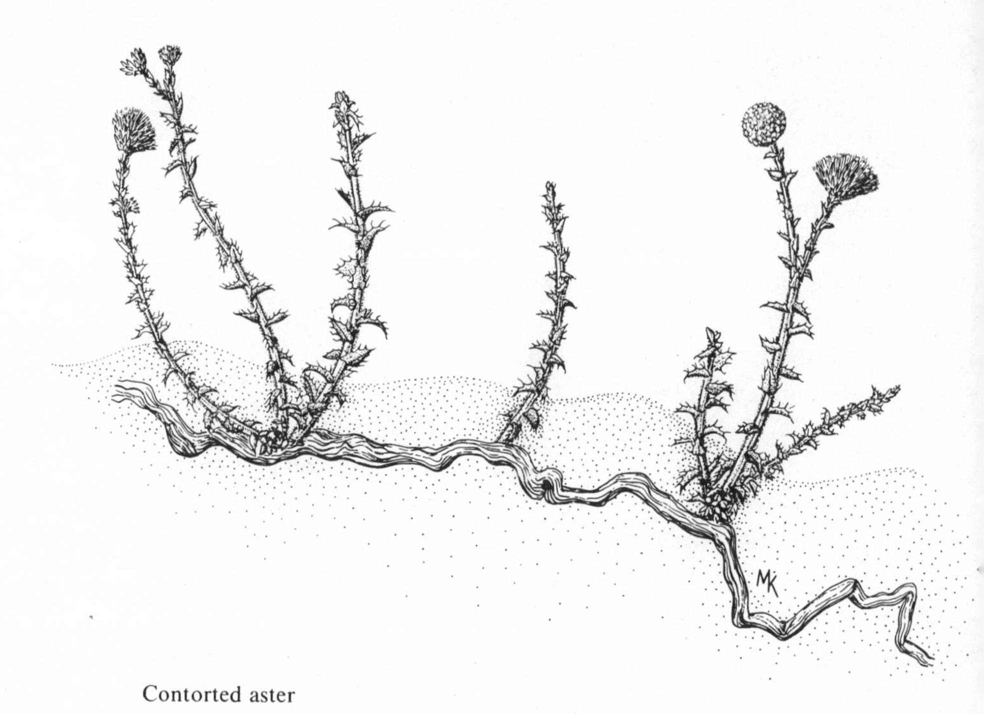

Contorted aster

off under the weight of the sand, and their leaves are too hot for photosynthesis in midsummer, even when the dunes are moist from summer rains. These plants are not well suited to the sand dune environment, and they eventually die without passing on their genes. In the meantime, the asters with more flexible stems and hairy leaves are thriving. As drifting sand covers their stems, they send out adventitious roots and leafy flowering stalks. Because of the coat of woolly hairs on their leaves and stems, they are able to photosynthesize when temperatures are high. They are well adapted to the sand dune habitat, so live long enough to pass along the traits that help them survive in active sand. We can readily see that contorted aster evolved as a result of the selective forces in its environment: moving sand, sunlight, gypsum, and perhaps others not so easily identified.

Similar forces have been at work over thousands of years on other dune fields in the Southwest. The Monahans Sand-

Peirson's locoweed

hills in western Texas support a shin- to waist-high woodland of shinnery oak, a leathery-leaved oak endemic to sand dunes from western Texas to eastern New Mexico. Its slender stems grow from stout rhizomes buried four to eight inches below the surface of the sand. Shinnery oak, like contorted aster, evolved to fit its environment, and its creeping rhizomes and large seeds suit it admirably for colonizing and stabilizing loose sand.

Other dune endemics in the Chihuahuan Desert are sand beargrass, plains yucca, dune devil's claw, and Buckley's penstemon.

Sonoran Desert dune fields have their share of endemic plant species, too. Peirson's locoweed, with its lavender flowers and silky leaves, is one of the rarer endemic plants on the Algodones Dunes. Like other dune endemics, it is precisely adapted to its environment. The inflated seed pods blow readily across the sand, collecting in swales and at the base of shrubs. Its seeds are larger and heavier than those of any other locoweed in the Southwest and thus can germinate and emerge even when buried in several inches of sand. After germination, the root plunges deeply into the sand, seeking moisture for the seedling. A young plant only a foot or so tall may have a taproot that extends six feet or more straight down into the sand before branching into lateral roots. Although Peirson's locoweed matures into a small shrub, plants may bear flowers and fruits just two months after germination, ensuring survival of the species even if the young plants do not live to adulthood.

Some endemic plants on the Algodones Dunes evolved under unusual circumstances. During the Pleistocene, the Gulf of California advanced and retreated several times. The present Gulf of California lies nearly one hundred miles south of the Algodones Dunes, but in the Pleistocene it may have extended at times to where the southern tip of the dune field now lies. As the Gulf retreated and advanced, dune plants were separated into coastal and inland populations. Natural selection operated on both populations, producing inland species closely related to but distinct from those on the coast; thus, a close relative of Peirson's locoweed—silvery locoweed—is endemic to dunes along the northern Gulf coast. There are other inland-coastal dune species pairs, as well. Wiggins' croton grows on inland dunes in the Sonoran Desert, and the closely related California croton grows on coastal dunes. Dune sunflower grows on inland dunes, silvery sunflower on coastal dunes. Other endemic plants on Sonoran Desert dunes include giant Spanish needles, dune buckwheat, and dune spectacle-pod.

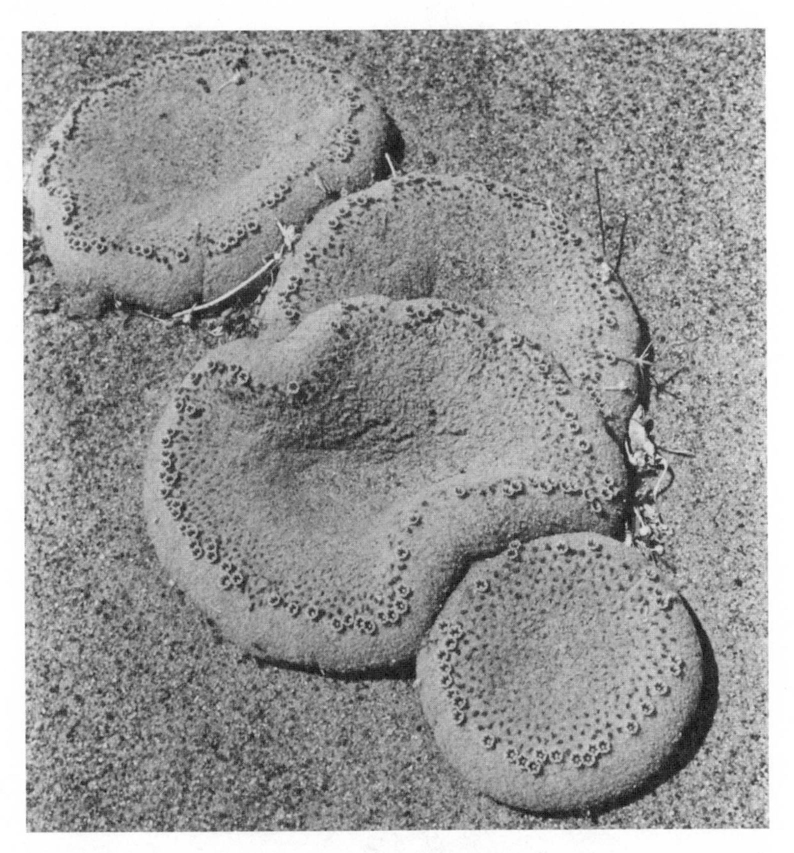

The flowering heads of sand food lie right on the sand.

One of the most interesting endemic plants on dunes is sand food, a fleshy parasite that grows on the Algodones and Gran Desierto dunes. The flowering head of sand food is the size and shape of an overgrown mushroom. It appears on the dunes between March and May, when the surface of the sand is very hot—140° F or more. A dense coat of short, gray hairs protects the head from the heat and reflected sunlight of the sand. Although the flowering head apparently lacks any sort of stalk, it actually grows from a thick, leafless stem that curves as much as five feet down into the sand to its point of attachment on a host plant.

Since sand food lacks chlorophyll and cannot manufacture

Long-leaf sunflower

the sugars it needs for growth, it steals sugar from the host plants it parasitizes. Sand food has several known hosts, including white bursage, dune buckwheat, pleated coldenia, and dune peabush. No one knows how sand food seeds locate an appropriate host. Some think that kangaroo rats bury the seed heads, unintentionally promoting contact between seed and host. Others believe that the dried seed heads break off the parent plant and tumble across the dunes in the wind, dispers-

Welsh's milkweed grows nowhere but the Coral Pink Dunes.

ing seeds as they go. Some of these seeds could be trapped in the base of potential host plants and buried by drifting sand. Of the thousands of seeds produced by a sand food plant every year, only a few germinate and become established.

Plants endemic to dunes in the Great Basin Desert include long-leaf sunflower and Welsh's milkweed. One of the dominant plants on the Coral Pink Dunes, long-leaf sunflower also grows on other dunes on the Colorado Plateau. Its cycle of growth is well suited to the patterns of sand movement on these dunes. The flowering stems grow just rapidly enough to outstrip sand accumulation throughout the summer. Stems and roots catch blowing sand, building hummocks under the plants. At the end of the growing season, the leafy stems disintegrate: some break off and blow away, others remain stuck in place like broomstraws. Snow occasionally blankets the dunes during the winter, and both plant growth and sand movement stop temporarily. In the spring new stems arise from the intricately branched roots, and by August, floppy, over-

Dune grass

sized sunflowers cover every hummock.

Welsh's milkweed, which grows nowhere but the Coral Pink Dunes, is less common there than long-leaf sunflower. It grows from perennial roots in the summertime and dies back to the ground every winter. The leaves are flocked like flannel with matted hairs, giving the plant a silvery sheen. The hairs probably help cut water loss and may keep the leaves, particularly young, sensitive ones, from being too hot to carry on photosynthesis in the summer.

Dune endemics in the Mojave Desert include dune evening primrose, sand locoweed, Borrego locoweed, and dune grass. Dune evening primrose, like dune grass and sand locoweed, is endemic to the Eureka Dunes. In summer, fall, and winter, dune evening primrose plants are rosettes, that is, clusters of leaves that lie flat on the sand. In spring, the rosettes send out leafy shoots up to two feet long, each tipped with a white flower that blooms at night and collapses like wet tissue paper by the following day. Seeds disperse during the early summer, then drifting sand covers the stems. In the fall, a new rosette of leaves grows at the tip of each stem. Each rosette becomes a new plant the following spring. This unusual growth pattern is particularly well suited to the season of the wind at the Eureka Dunes, since the plants grow fastest when the sand is most active.

As we have seen, many dune fields in the Southwest support at least one endemic plant species. Some have more. In contrast, most nondune desert areas of similar size have few endemics; many desert mountain ranges in the Southwest have none at all. Why do dune fields have more endemic species than the valley floors right next to them or the desert mountain ranges just five or ten miles away? First, moving sand has been a powerful force in the evolution of plants that are specially adapted to the sand dune environment. Comparable selective forces do not exist on most valley floors or desert mountain ranges. Second, sand dunes are islands in a sea of desert. There is too much inhospitable territory between dune fields for the seeds of most species to disperse from one to another, so species remain isolated on their own particular islands of sand.

Endemic plant species are one of the most interesting aspects of sand dune biology, and one of the most irreplaceable, as well. Sand dune endemics are the product of thousands of years of interaction between a particular gene pool and the environment. Although the environment may be duplicated in other parts of the world, the plants themselves never will.

Endemic plants are unique,and because their numbers are relatively few and their distribution is restricted, they are particularly vulnerable to extinction. A widespread plant such as creosote bush, which occurs in large populations throughout much of the desert Southwest, is relatively insensitive as a species to minor changes in the environment. The gene pool of creosote is so large and varied, there is probably a genotype that could survive any particular environmental upheaval. But for the small populations of dune endemics, no such large and varied gene pool exists. Greatly decreased rainfall, greatly increased temperatures, frequent severe freezes: these or other changes could wreak havoc on populations of dune endemics, which might not have enough genetic variability to adjust.

Nature is not the only cause of environmental change. People change the environment, too, and dunes are likely targets for change because of their recreational potential. When a dune field is open to off-road vehicles, plants in heavily used areas are destroyed outright. The remainder survive in pockets where traffic is light. In effect, populations of dune plants, already restricted in distribution, are restricted still further. The total number of plants in the population is decreased, thereby decreasing its genetic variability. The diminished population becomes less able to adapt to environmental changes. When the population reaches a critically small size, extinction is inevitable, and extinction, as we well know, is forever.

Sea Sands

THE TIDE WAS OUT AS WE WALKED ALONG THE Gulf coast near Desemboque, Sonora, one morning. Two hundred yards of rippled mud separated the first dunes from the line of breakers that curled off the rim of the ocean. Pools of sea water on the tideflats glared at the sky. Near one pool a lone gull sat as still as a bleach bottle, then flapped away at our approach. Walking along the tideline, we found glass bottles, still unbroken; strands of rope bleached whiter than the sand; disarticulated bird bones and matted feathers; a cow's horn; tangled fishing nets; and shells, most broken, but a few still whole and with the bloom of the ocean yet upon them.

Shells were also plentiful amid the rolling dunes that stretched for miles along the coast south of Desemboque. Some of these shells had been neatly pierced with circular holes; others had been worked along the edges, as though to make tools for scraping. With the shells we found broken shards of pottery and an occasional grinding stone, just the size to fit in one's hand. It was easy to imagine Indians making camp in the dunes a thousand years ago. They came to the

Active dunes at Desemboque on the Gulf of California

coast to harvest cockles, conches, and clams from the mudflats. Probably they took basketfuls of shellfish back to the dunes, where firewood was plentiful. Periodically, they could have found fresh water for cooking and drinking in the Rio Asuncion, several miles south of Desemboque.

Later in the day we went to see the river for ourselves. Like most desert streams, the river is dry except after rains, when it flows bank to bank for a short time, carrying a heavy load of sand and silt from the mountains that lie inland. When we saw it, just enough water was flowing to make a hundred anastomosing stream channels, not one more than an inch or two deep. Far away from where we stood on the riverbank we could see the ocean's waves curling into the mouth of the river. We had heard that sometimes dunes blocked the flow of the river to the sea, but the channel was free of dunes that day. There was no lack of sand on the river's broad floodplain, however; there was sand enough to build a river of dunes.

Dunes on the coast are built from sand carried down to the sea by streams and rivers, and the largest coastal dune fields often lie near the mouths of rivers, as at Desemboque, where sand from the Rio Asuncion has been blown into dunes. In most years the rivers in the desert do not reach the sea; only the rare flood waters tumble into the Pacific Ocean. The average yearly rainfall at Desemboque, and along much of the desert coast, is about four inches. Such low rainfall results in low rates of erosion. Scientists have estimated that the greatest quantities of sand are produced when rainfall is about twelve inches per year. Presumably, greater amounts of rainfall nourish enough vegetation to hold the soil in place, and lesser amounts result in little wearing of rock into sand-sized fragments. Because of the low rainfall and the sporadic flow of the rivers along the coast at present, it seems likely that many of the desert dunes on the coast were built from ancient deposits of sand transported to the coast thousands of years ago when the climate was wetter.

Baja California and Sonora together have about 2,700 miles of coastline. About two-thirds of this coast is desert and perhaps one-third of the desert coast is sandy. Some of the major dune fields along the coast of Baja California occur at Sebastian Vizcaino Bay, San Ignacio Lagoon, and Magdalena Island. On the desert coast of Sonora, in addition to the dune field at Desemboque, there are dune fields near Punta Cirio, from Kino Bay to Tastiota Estuary, near Rocky Point, and around El Golfo along the western edge of the Gran Desierto.

The topography of the Pacific and Gulf coasts is such that dunes typically form parallel to the coast and at right angles to the prevailing wind. Closest to the beach we find low, rounded, hummock dunes called foredunes, which are created as creeping, shallowly rooted plants catch and hold blowing sand. Behind the foredunes we find one or more rows of larger dunes, walls of sand up to forty feet high, which may front the coast for miles. In geomorphic terminology, these are transverse dunes, but because of their position along the coast, they are often called backdunes.

Plants growing on coastal desert dunes face all the problems known to plants on inland desert dunes, and a few more. High temperatures, a long dry season, moving sand, and low levels of necessary nutrients are just as characteristic of coastal desert dunes as of inland desert dunes. Further hazards of coastal dunes are the saline water table, which lies just below the ground along the coast, and the desiccating winds that blow continually. In addition, the sand of coastal dunes can be very salty because of constant salt spray and occasional inundation by sea water.

High concentrations of salt in soil or water inhibit plants from taking water out of the soil. Water follows what plant physiologists call a water potential gradient: it flows from regions of high water potential to regions of low water potential. Under ordinary conditions in nature, the water potential inside plants is lower than that in the soil, so that water flows from the soil into the plant. Salty soil, however, has a much lower water potential than normal soil. Since the water potential of salty soil is low, plants cannot pull water from it as readily as from normal soil, and they die or suffer salt damage.

Plants that can grow in salty soil are called halophytes, that is, salt-plants. Levels of salinity that would kill most plants are part of the halophyte's daily environment. How do halophytes tolerate saline conditions that damage or kill most other plants? Some are able to accumulate larger and larger quantities of salt in their tissues throughout the growing season so that their water potential is always lower than that of the salty water in the soil. Such plants maintain an even, rather than an increasing, concentration of salt and have succulent leaves or stems. Other halophytes regulate salt concentration inside the plant by exuding salt through special glands on their leaves.

Both of these adaptations to salinity are found among the various coastal dune plants. Seaside sand verbena has fleshy leaves, and the various saltbushes are encrusted with salt deposits. Several dune plants found along the arid stretches of the coast typically grow on saline playas farther inland: these include pickleweed, salt grass, seepweed, and frankenia, among others. The same adaptations that enable them to live and even thrive where salt coats a playa with a crunchy layer of white crystals also fit them for survival in the salty sand and ocean

Seaside sand verbena

spray along the coast.

Seaside sand verbena, which grows on dunes along the Pacific coast from Morro Bay, California, to the tip of Baja California, and along both sides of the Gulf coast as far south as Sinaloa, is a good example of a plant that is ideally suited to the coastal dune environment. A characteristic species of foredunes, seaside sand verbena tolerates both salt and sand movement and is an important builder of foredunes. The woody taproot is profusely branched, and as its many creeping stems are buried by sand, they continue to grow at the tips and send out side branches, forming dense mats. The mats catch blowing sand, thus building a mound under the growing plant. At the same time, its prostrate growth form lets it avoid salt spray. Leaves of seaside sand verbena are succulent inside and leathery outside. Succulent leaves enable it to tolerate high salinity, and the leathery exterior prevents undue desiccation by coastal winds.

The roots of seaside sand verbena do not penetrate deeply into the sand. At the coast, two water tables lie one above the other: a shallow layer of fresh water on top of a thicker layer of salt water. Since plants with deep roots would tap not only fresh water but unusable salt water, the shallow roots of seaside sand verbena are an important adaptation to its coastal habitat.

Seaside sand verbena is precisely adapted to its coastal environment throughout its life cycle; it even depends on the ocean for dispersal. Seaside sand verbena seeds are chunky compared with the seeds of inland sand verbenas, whose big wings and light seeds ensure that they will be blown long distances by the wind, at least long enough to hop from one dune island to another. Because seeds of seaside sand verbena lack wings and are heavy for their size, they are not at all suited for wind dispersal. However, the large air pockets in the spongy seed coat make the seed buoyant, and the thickness of the seed coat protects the delicate embryo inside from salt damage. Adaptation for ocean dispersal is one reason that seaside sand verbena is so widespread along the Pacific and Gulf coasts.

Silverleaf saltbush, which grows along the Pacific coast from northern California to central Baja California, is another dune plant that is nicely adapted to the coastal environment. Like seaside sand verbena, silverleaf saltbush grows on foredune hummocks as close to the ocean as plants can grow. Also like seaside sand verbena and other foredune plants, silverleaf saltbush grows low to the ground, building hummocks as sand becomes trapped among the leaves and stems. Its seeds are also enveloped in a thick, spongy coat that contains air pockets.

A dense coat of tiny hairs, like collapsed or exploded balloons, covers the leaves of silverleaf saltbush. These little balloons, which are typical of saltbushes, give the plants their silvery appearance and are crucial in helping them survive in saline habitats. Salt fills the balloon hairs, reaching concentrations up to sixty times greater than that of the leaves. Eventually the hairs burst, coating the leaves with salt. The leaf hairs also lower the surface temperature of the leaf, keeping it closer to the best temperatures for photosynthesis.

Vegetation patterns on coastal dunes reflect increasing salinity and sand movement as we move from backdunes to the shore. Salt spray blown inland prunes plants that grow near the ocean, shearing them close to the ground. The closer we get to the sea, the thicker the salt spray becomes and the shorter the vegetation grows; shrubs that are ordinarily waist-high reach only up to our knees, and normally knee-high plants are no taller than our shins. At Punta Cirio, even the ponderous cardon cactus, with its massive trunk and candelabra of stout arms, shows the combined effects of salt spray and wind, dwindling from perhaps forty feet inland to twenty feet near the beach.

A wider variety of plants grow on backdunes than on foredunes, a reflection of lesser salinity and sand movement just ten or twenty yards from the beach. One typical backdune plant is California croton. Closely related to Wiggins' croton, California croton is a sprawling, mound-forming plant, rather than a single-trunked, upright shrub like its inland relative. While Wiggins' croton anchors itself with a few vinelike roots, California croton branches profusely above and below the ground, so that roots and stems form mounds that catch and hold blowing sand. California croton tends to grow most abundantly on the lee side of the coastal dunes, probably because it is not very salt tolerant. Other sprawling perennials that grow on backdunes include silvery sunflower, various species of brittlebush, dune peabush, silver spurge, silvery locoweed, Barclay's saltbush, and frankenia.

How do coastal desert dunes compare with saline dunes in the inland deserts? From the limited information available, it appears that inland dunes can be as salty as coastal dunes. Not all inland dunes are saline, of course, and even in those that are, salinity varies a good deal from place to place. Since most dune-adapted species cannot tolerate high salinity, halophytes from the margins of nearby salt flats take their place on inland saline dunes. Salt grass, greasewood, and fourwing saltbush, all salt-tolerant plants often found around dry lakes, are typical of saline dunes, although they are not dune plants in a conventional sense. Other salt-tolerant species that will grow on inland dunes include pickleweed, sand dropseed, and seepweed.

We have seen that saline dunes are populated by halophytes from nearby playas and that coastal dunes are populated by seeds that drift with the ocean's currents. But how are other sand dunes populated? How do plants move from one dune field to another?

Stragglers and Stepping Stones

WE STUMBLED UPON A DUNE FIELD BY ACCIDENT one spring as we drove back roads on the Colorado Plateau. We wanted to go from Pipe Spring National Monument in northern Arizona to Nelson, an old mining town in southern Nevada, and although it is only 145 miles from one to the other as the crow flies, the crow may fly over the Grand Canyon, but we could not. Heading north, we drove through the village of Cane Beds, Arizona, then crossed the state line into Utah. Suddenly we found ourselves in slick rock country, red rock country, spectacular country where ramps of scalloped sandstone are capped by massive rock walls. We did not know it was dune country, too, until we glimpsed a hillock of pumpkin-colored sand rising like a harvest moon over junipers to the east.

These dunes, the Coral Pink Dunes, lie at 6,000 feet in the mouth of Sand Canyon, a flat-bottomed, steep-walled canyon that cuts down through the southern edge of the Wygaret Terrace. The dune field is small, only about seven square miles in extent. The highest dunes, in the center of the dune field, are rolling transverse dunes about seventy-five feet tall. Lower parabolic dunes surround the active ones. Since snow and frost anchor the active dunes during the winter, their movement is slow, only a foot or two each year.

The Coral Pink Dunes were built from twice-recycled sand. In the age of dinosaurs, two hundred million years ago, dunes covered a vast desert in what is now northeastern Arizona, northwestern New Mexico, and southern Utah. With the passage of time, sand grains in the dunes were cemented by calcite and iron oxides into the massive, scalloped beds of striated rock so characteristic of slick rock country. In some places, the rock—known as Navajo sandstone—is white; in others, iron hematite colors it red or pink. Because the cement between the sand grains is weak, Navajo sandstone is friable and weathers readily into loose, red sand. This sand is blown into small, active dune fields such as that in Sand Canyon and into broad, stable sand sheets such as the one that covers over 25,000 square miles in northeastern Arizona and adjacent New Mexico.

For those interested in plants, summer is the time to see the Coral Pink Dunes, for summer is when they burst into bloom. Mounds of long-leaf sunflower sag under their burden of blossoms: yellow flowers nodding among the leaves like teacups hanging from cuphooks. Clusters of wild buckwheat and silvery sophora dot the coral-colored sand, a kaleidoscope of pink and white and blue.

One summer afternoon, we wandered over the dunes, collecting a few plants to identify later and listing those whose names we already knew. Our list of some three dozen names was just a partial list of the plants that grow on the Coral Pink Dunes, reflecting those we saw in one part of the dune field at one time of year. Botanists often compile plant lists for large areas, returning many times throughout the growing season over a period of years. The final product, a complete list of the plants of the area, is called a flora. (It is important here to note the difference between flora and vegetation. Flora refers collectively to all the plant species in an area, whereas vegetation

A wide variety of plants grows on the Coral Pink Dunes.

refers to the abundance, height, and density of the characteristic plants.)

Literally hundreds of floras have been compiled for different places in the Southwest. Many national parks and monuments have floras: Organ Pipe Cactus National Monument, Zion National Park, Bryce Canyon National Park, and Rocky Mountain National Park, to name a few. Botanists have also compiled floras for many of the larger mountain ranges in the Southwest, including the White Mountains in southeastern California, the Charleston Mountains in southern Nevada, and the Sandia Mountains in northern New Mexico. Dune fields, however, have been given short shrift for many years; it is only recently that botanists have begun to inventory the plants on southwestern desert dunes. So far, floras have been compiled for White Sands National Monument, the Algodones Dunes, the Coral Pink Dunes, the Kelso Dunes, the Eureka Dunes, and the Gran Desierto dune field. Natural resource specialists use floras to make management decisions. Scientists use floras to examine the distribution of species and to follow immigration and extinction of species in an area.

Using the floras of desert dune fields, we can explore some of the fundamental problems of plant geography. One of the basic questions in plant geography is: how did these plants get here? We can take this question further and ask why we find some plants only on particular dune fields. That leads us to wonder how similar the floras of different dune fields are and why.

A large proportion of the plants on every dune field are typical desert plants. They are widespread in nondune habitats, and on dune fields, grow strictly on marginal dunes, where the sand is stable enough that they are not threatened by immediate burial or excavation. Such stragglers are typical of stable dune fields throughout the Southwest. On the Coral Pink Dunes, some of the stragglers are big sagebrush, Gambel oak, Mormon tea, and bitterbrush. On the Kelso Dunes, the stragglers include creosote bush, white bursage, catclaw, and armed senna. It is not difficult to envision how stragglers arrived on dunes: seeds from nearby plant communities landed on the sand, and a few germinated and became established.

For some plants, we can make a case-by-case examination of distribution. Thurber's penstemon, which grows on the Kelso Dunes, is not a typical dune plant, nor is it characteristic of the desert flats nearby. However, since this penstemon is a favorite browse of cattle, it seems likely that cattle, which range freely over the Kelso Dunes, introduced it. Cattle may also have brought cobweb phlox onto the Kelso Dunes. It, too, is unknown from any other desert dune field and is relished by livestock in spite of its prickly leaves.

Certain plants, particularly annuals, grow on many southwestern dune fields. Often these plants are adapted for wind dispersal and pop up wherever there is a suitable habitat. For such plants as hoary dicoria, ringwing, palafoxia, sand verbena, spectacle-pod, Spanish needles, and Riddell's groundsel, dunes are islands of sand in an ocean of desert, and pockets of shallow sand between the large dune fields are stepping stones across the ocean.

What about sand dune plants that have no particular adaptation for dispersal? How do they get from one dune field to another? Dune peabush, which grows on both the Algodones and Gran Desierto dune fields, is one example. Others are broom pea, sand sage, mesa dropseed, giant dropseed, plains penstemon, hooded spurge, and Parry's spurge, all of which grow on both the Samalayuca Dunes and the Monahans Sandhills. Here, too, the key seems to be pockets of sand that link major dune fields. Sandy plains west, east, and south of Yuma, Arizona, might have acted as stepping stones between the Algodones and Gran Desierto dunes; the dunes near El Paso, Texas, might have been the stepping stones that linked the Samalayuca and Monahans dune fields.

Why do some plants grow only on one or two dune fields? In most cases, these species evolved in place and never spread to other dune fields. Dune grass on the Eureka Dunes is probably such a plant, as well as giant Spanish needles on the Algodones Dunes and Welsh's milkweed on the Coral Pink Dunes. Of course, it is always possible that further exploration may find some of these plants on nearby dune fields. Then, too, it is possible that their present distribution is only a fraction of what it once was. But if we assume that they evolved in place and never spread to other dune fields, then we are back to the original question.

Perhaps we should instead ask *why* these endemic plants never spread beyond their "birthplaces." One reason is that the pockets of shallow sand that act as stepping stones for some dune plants will not suit those adapted to active, deep sand. Another reason is that the birthplaces of some endemics are extremely isolated, separated from the nearest dune field by miles of rocky desert or by high mountain ranges. Time could be another reason. Moving from one place to another is a slow process for plants with no particular adaptations for long-distance dispersal, and given enough time, some dune endemics may yet spread to nearby dune fields.

The distribution of dune plants, whether widespread or restricted, leads to the question: how similar are the floras of different dune fields? On some Old World dune fields, as in the Kalahari Desert in southern Africa, the flora is a cohesive unit. Although some variation occurs from north to south due to higher rainfall in the north, most of the 350-odd plant species occur throughout the entire 62,000 square miles of the dune field. In other words, there is a group of plants characteristic of sand dunes in the Kalahari Desert, a sand dune flora.

Is there a sand dune flora in the Southwest? Do the species on the many different dune fields form a cohesive unit? A similarity analysis might give us the answer.

If we add the number of species on, say, the Algodones Dunes, to the number of species on the Kelso Dunes, and then divide that figure into the number of species shared by the two dune fields, we come up with a similarity index, a number that indicates how similar or dissimilar the two floras are. The closer the similarity index is to 1, the more similar are the floras; the closer it is to 0, the less similar they are. The similarity index for the Kelso-Algodones dunes floras is 0.324. Apparently the floras of these two areas do not have much in common. The similarity index for the Algodones–Gran Desierto dune floras is 0.512, a rather high value. The Coral Pink–Kelso dune floras have a similarity index of only 0.050, a low value. The White Sands flora is quite dissimilar to all the others: the White Sands–Kelso dune flora pair has a similarity index of 0.012; White Sands–Gran Desierto has a value of 0.044; White Sands–Algodones, 0.025; and White Sands–Coral Pink, 0.123. Evidently the floras of most southwestern dune fields are more dissimilar than they are alike. No single group of plants characterizes all of the dune fields in the Southwest.

Why do such apparently similar habitats share so few species? What barriers prevent plants that are thriving on one dune field from growing on another?

One of the most important barriers is climate. Dune fields in the Chihuahuan Desert receive summer rains, those in the Mojave Desert, winter rains, and those in the Sonoran Desert, both summer and winter rains. There are also important differences in summer and winter temperatures among these different deserts. It is unlikely that plants adapted to one climate will thrive under another.

Substrate is another barrier to widespread distribution of dune plants in the Southwest. The gypsum sand of White Sands lowers its similarity to other dune floras. Species adapted to gypsum may thrive on the White Sands dune field, but many cannot live on quartz sand. Conversely, many plants that grow on quartz sand cannot tolerate gypsum. Salinity also promotes dissimilarity between dune fields. Halophytes such as pickleweed, seepweed, and salt grass grow on the Death Valley dunes but not on the Algodones, Kelso, Eureka, Gran Desierto, or Coral Pink dune fields.

The stragglers that grow on the margins of southwestern dune fields are not the same from place to place, and that is another reason for the marked dissimilarity in dune floras. Largely because big sagebrush grows on the valley floor next to the Coral Pink Dunes, it also grows on the periphery of the dunes themselves. But big sagebrush does not grow on the Kelso or the Eureka or any of the other dune fields discussed. In the same way, prince's plume grows on and near the Eureka Dunes, and armed senna on and near the Kelso Dunes, but neither grows on the Coral Pink Dunes.

Additional factors that contribute to the lack of similarity between dune floras in the Southwest have already been discussed: these include isolation, lack of adaptation for long-distance dispersal, and lack of time for dispersal from one dune field to another.

In sum, the floras of southwestern dune fields are not as much alike as we might expect them to be. Species do not move freely from one dune field to another. There are too many barriers, for one thing—gypsum sand, inimical climates, high mountain ranges—and most dune endemics are not well adapted for long-distance dispersal.

Dune floras are a slice in time. Some species are on their

way out, some on their way in, some static. Dune floras are a slice in space, too. Many species on any dune field are there simply because they grow nearby. But others, the dune endemics, are there because sand dunes are the only places where they can live.

Other Deserts, Other Dunes

I ONCE SAW OLD WORLD DUNES DURING A plane flight across the Sahara from Algiers to Niamey. For hours the plane hung above a carpet of sand; only an occasional rocky plain or hill poked a hole in the seamless, red fabric. Huge as it was, this sand sea was just one of many in the Sahara. The Grand Erg Occidental, a vast desert of sandy plains and dunes that lies south of the Atlas Mountains and west of the Morocco-Algeria border, covers 64,000 square miles, an area roughly the size of the state of Washington. Farther east, the Grand Erg Oriental covers 119,000 square miles, about seventy percent of which is sand. One sand sea in Libya is said to be nearly 4,000 feet thick, as deep as some mountains are high.

Other major sand seas of the world occur in the Namib in southeast Africa, the Kalahari Desert in southern Africa, the Arabian Peninsula, the Thar Desert in India, and the Takla Makan Desert in China. The largest of these covers 162,000 square miles; the smallest, 21,000 square miles. Our southwestern dune fields would be lost amid these vast piles of sand. Even the 1,700 square miles of sand dunes in the Gran Desierto, the largest sand sea in North America, seems tiny in comparison to sand seas in Old World deserts—more like a sand lake than a sand sea—while still smaller dune fields such as the Coral Pink or Eureka dunes are mere sand puddles. How else do dunes in the arid Southwest differ from those in other deserts? How are they alike?

Dunes in the Namibian Desert are mostly long sand ridges running parallel to the Atlantic coast for miles. These huge sand ridges, up to three hundred feet tall, are separated by equally long valleys about a mile wide. The Namibian Desert is extremely arid: rainfall is so rare and erratic that annual averages mean little in terms of plant and animal life. Once every forty or fifty years a veritable deluge of rain might fall, as much as five inches over three months. But more typically, year after year goes by with little measurable rainfall. Droughts of thirteen years or more in which yearly rainfall averages less than half an inch are the rule rather than the exception.

That any plants at all can survive in such a climate seems something of a miracle, and if not for the occasional fogs, perhaps none could. Close to the coast, fog occurs on about 120 days in the year; farther inland, about one-third as often. Fog is more reliable than rain in the Namibian Desert, and in any month, fog is three times more likely than rain. The effects of fog are very real: fog droplets condense on plants and sand in measurable quantities, from one-half to one and one-half inches per year. In many years, moisture from fog is more abundant than rainfall.

Most of the time, plants are not abundant on the Namibian dunes. During the frequent (and normal) prolonged droughts, only two species are able to survive: one a perennial grass, the other a succulent perennial related to the ice plants so common on the California coast. The adaptations of the succulent, whose scientific name is *Trianthema hereroensis,* to its harsh environment are particularly interesting. Since it can bear flowers and fruits throughout the long dry periods, it appears to be independent of moisture. In reality, *Trianthema here-*

roensis is one of the few plants known to absorb fog through its leaves. For many plants, succulence is an adaptation to salinity, as we have seen. But for *Trianthema hereroensis,* succulence is an adaptation to drought. Empty spaces permeate the tissue of each leaf, and water stored in these spaces makes the leaves succulent. Droplets of fog that condense on the leaves are absorbed and diffused rapidly throughout the plant, reaching the roots in a matter of hours. In addition, heavy fog condenses on the plant and drips onto the sand, making water directly available to the roots.

Other plants on the Namibian dunes have adapted to drought by escaping it. Two species of grass germinate abundantly on the dunes when there is a series of relatively wet months. These grasses persist for several years, then die when no more rain comes. Botanists call such plants quasi-ephemerals. The remainder of the species that grow on these dunes are true ephemerals: they germinate only after sufficient rain has fallen and die soon after they go to seed. Pulses of rainfall thus cause a burst of plant growth on the dunes, but it is over quickly, leaving the dunes as bare as before.

No dune area in the Southwest has so arid a climate as the Namibian dunes. Although the Gran Desierto is the most arid region in North America, it is well watered compared to the Namibian Desert: three inches of rainfall annually as opposed to one-half inch. The dozens of perennial plant species that live on the Gran Desierto dunes are an emphatic contrast to the two that inhabit the Namibian dunes.

Apparently, little is known about the effect of long periods of drought on dune plants in the Southwest. We can imagine, though, that the first plants to feel the effects of drought would be seedlings. Because their roots are too short to reach water stored deep within the dunes, they depend on periodic rainfall to moisten the upper layers of dune sand. Shallowly rooted plants would succumb to drought after the seedlings, and finally perhaps even the most deeply rooted plants would exhaust the supplies of accessible water and die.

The Kalahari Desert lies inland from the Namibian Desert

in southern Africa. Like the Namibian dunes, those in the Kalahari are long, parallel ridges of sand up to one hundred feet high and one-third mile apart. Their flattened crests are often twenty to thirty feet wide. Rainfall in the southern Kalahari, although low, is less erratic than in Namibia. Annual rainfall averages between six and eight inches; about one-quarter of this falls in the winter, and the remainder comes in summer thunderstorms.

Unlike the Namibian dunes, where vegetation is sparse except after rare, heavy rainfall, vegetation on the Kalahari dunes is relatively dense. Small trees and perennial grasses form an open savanna on the slopes, which are largely stable. The crests, however, are active and mostly barren. This pattern is maintained by a bottleneck in seedling establishment. Few seedlings become established on the loose sand of the crests due to constant sand movement, extremes in sand temperature, and rapid desiccation of the surface after rainfall. Seedlings are much more likely to thrive when they grow among the trees and grasses already well established on the dune slopes. Thus is the pattern perpetuated as the slopes continue to support trees and grasses, and the crests continue to be mostly bare.

Why do dunes in the Kalahari support trees when trees are so rare on desert dunes in the Southwest? Why does the active sand on the crests not overwhelm the trees on the dune slopes and bury them, as occurs on dune fields in our area? We find the answers to these questions in the morphology of the dunes themselves. The parallel dunes typical of the Kalahari are maintained by prevailing winds that blow along the long axis of the dunes. Loose sand drifts downwind along the crest, and since it mostly stays right on the crest, plants on the stable slopes are in no danger of being engulfed by moving sand.

Plants on the Kalahari dunes have various adaptations to drought. A few grow from bulbs that have succulent roots. In periods of drought, the root tissues contract, pulling the bulb deeper into the sand where water might still be available. Another adaptation is deciduous roots, seen in plants with the long, sparsely branched roots typical of plants on active dunes. Mature roots of these plants are corky on the outside, an adaptation that helps prevent undue water loss. When the sand is wet from recent rainfall, many tiny, cork-free roots are formed along the main roots. Since these smaller roots are not covered with cork, they can rapidly and efficiently absorb the extra moisture in the sand, and when the sand dries out, they slough off with no harm done to the main root system. As on southwestern dunes, many grasses on the Kalahari dunes form sand-grain sheaths around their roots.

The Simpson Desert in south-central Australia covers about 73,300 square miles, much of it sandy. Rainfall is less than seven inches per year in the more arid parts. Extended droughts alternate with wet periods lasting several years. Parallel dunes a quarter of a mile apart and up to one hundred feet high are the most common type of dune in the Simpson Desert.

As in the Kalahari, the dune slopes are stable, and the crests are active. Scattered hummocks of cane grass grow on the crests of the dunes along with a few other grasses, herbs, and woody perennials. The vegetation on the slopes is an open savanna much like that on the Kalahari dunes. But unlike the Kalahari dunes, those in the Simpson Desert are swept by periodic wildfires, which perpetuate the pattern of vegetation on the dunes.

In general, fires are not common in true deserts, where there is seldom enough fuel to allow wildfires to burn over large areas. On the Simpson Desert dunes, however, a series of wet years provides fuel in the form of plentiful grasses and other herbs. After plants of the lower dune slopes and swales burn, many grow back rather quickly. Such fire-tolerant species have adapted to fire in various ways: a common one is sprouting from the roots after the top of the plant has burned to the ground. These same species, however, cannot tolerate the low nitrogen and moving sand of the active crests and so remain on the slopes and swales. The grasses and other plants that thrive on the crests have adapted to moving sand and low nitrogen, but fire kills them outright. They must reproduce from seed after fire. When fire destroys the plant cover on the crests, the sand becomes even more active. Thus fire keeps the dune crests unstable and prevents shrubs and trees from becoming established except on the slopes. On most desert dunes in the Southwest, the vegetation is too sparse to allow fire to spread. Perhaps wildfires could sweep over stabilized dunes where big galleta grass grows in dense stands, but they seldom, if ever, do.

One of several sand seas on the Arabian Peninsula is An Nafud in northern Saudi Arabia. Covering 45,000 square miles, An Nafud lies in a region that receives less than four inches of rain annually, and that little is erratic from place to place and year to year.

Although the greater part of the An Nafud sand sea is barren, scattered plants grow in some areas. On deep sand one finds *Calligonum comosum,* a five-foot-tall shrub with a stout trunk and long, whiplike roots. The needlelike leaves of *Calligonum comosum* are shed during the dry, hot summers. When rains start in the cool season, *Calligonum* puts out flowers and new leaves. The tiny, drought-deciduous leaves help prevent undue water loss throughout the year. Although *Calligonum* grows on sand dunes, a relatively moist habitat in deserts, it must still conserve moisture. Even the water stored in sand dunes cannot entirely ameliorate the effects of an arid climate.

In the An Nafud, ephemerals such as dune caltrop germinate in the swales after rains. The fruit of dune caltrop is round and flattened, like an aspirin tablet. Small, hard prickles on top of the dry fruit stick to feet or hooves or tires, which is how the seeds are dispersed. Although each spiny fruit contains up to ten seeds, only one germinates when the first rains fall. If the rain turns out to be inadequate, the seedling withers and dies. The next rain causes a second seed to germinate, and it, too, will die if the rain has been insufficient. One by one the seeds germinate until one of the seedlings grows to maturity or until the seed supply is exhausted. Since the seeds contain only limited stores of carbohydrate as an energy reserve, the amount of rain is critical. Too little rain will not supply enough water to nourish the seedling until its taproot can reach deeper water stored in the dunes. Although some of the seedlings may die because of insufficient rainfall, the chances are that one will eventually become established. This is dispersal in time carried to extremes, a strong testimony to the undependability and irregularity of rainfall in the harsh desert.

Dunes are by no means uniform from place to place, or even from year to year, since they exist under a wide variety of climates that change through time and space. In general, dunes are relatively moist habitats in deserts and relatively dry ones in wet climates. Even so, on desert dunes, the drier the climate, the sparser the vegetation. At one extreme are the Coral Pink Dunes (just on the border between desert and woodland) where ponderosa pines, normally forest trees, thrive until overwhelmed by sand. At the other extreme are the Namibian dunes where only two perennial plant species survive during extended droughts. But even the Namibian dunes flush with green on occasion; it is only a matter of time.

Although trees are rare on desert dunes in the Southwest, they are a characteristic feature of those in the Simpson and Kalahari deserts. In the Southwest, active dunes are often confined to the center of a dune field that is circled by stabilized dunes. But in the Kalahari and Simpson desert dune fields, individual dunes have stable and active portions. The flora of the 62,000-square-mile Kalahari dune field is rather uniform, but the floras of the seven-square-mile Coral Pink Dunes and the three-square-mile Eureka Dunes have few species in common, even though they are closer together than the northern and southern ends of the Kalahari sand field. Fog is crucial to dune plants in the Namib, and fire is a regular feature of Simpson Desert dunes, yet neither plays an important role on dunes in our deserts. Still, dunes are dunes, after all, and we are not surprised to find that both *Calligonum comosum* on the Arabian Peninsula and dune buckwheat on the Algodones

Dunes have vinelike taproots for seeking out deeply stored moisture, nor that cane grass in the Kalahari and dune grass on the Eureka Dunes catch and hold blowing sand, forming hummocks that protect the plant from prying winds. Such adaptations work on active dunes the world over, making dune fields as alike as they are different.

Use or Abuse

THE GRAN DESIERTO—THE GREAT DESERT—LIES in northwestern Sonora and covers 3,000 square miles. Much of the Gran Desierto is a sea of umber sand with jagged, rocky hills that emerge from the sand like ships. Although the hills are low—one could scramble to the top of any one of them and back down in a single day—they are called sierras. The most prominent of these, the Sierra del Rosario, floats on the sandy sea, dunes lapping at its hull. Other dunes like distant breakers rim the horizon.

The Gran Desierto is a vast and empty place. I went there in the early spring with two friends and saw no one else. We saw no travelers besides ourselves on the dirt roads that wind along the plains between the dunes: no woodcutters or miners, no homesteaders or hunters. The only footprints in the sand were those of coyote and jackrabbit. The only sound at night was the call of a great horned owl. The loudest noise during the day was wind blowing against our ears.

What a contrast to the Algodones Dunes, just across the border to the north. Highway 78, which crosses the dune field

Dunes surround the Sierra del Rosario in the Gran Desierto.

the short way, has become a boundary between two worlds, between a natural area closed to off-road vehicles and an unnatural area open to them. North of the highway, the dunes wear a cloak of vegetation—a tattered cloak, to be sure, with many rips and holes, but a cloak nonetheless—while to the south, the dunes are naked, stripped bare by thousands of vehicles year after year. On weekends, the roadside turnouts south of the highway are crammed more tightly with vehicles than a used-car lot: dune buggies, all-terrain cycles, jeeps, trailbikes, four-wheel-drive pickup trucks, sedans, motorhomes. The roar of motors is a constant din, and exhaust fumes hang in the air.

On every weekend from September through May, hundreds, even thousands, of people come to the Algodones Dunes. Day and night they drive up and down the dunes, careening in circles, spurting uphill and plunging down. Children in crash helmets sputter about on motorized tricycles. Teenage boys,

The Algodones Dunes Natural Area

shirtless and sunburned, ferry their girlfriends, in shorts and bikini tops, from one side of the dune field to the other. Wives and mothers sit on folding chairs in the shade of motorhomes and watch.

For the most part, these are ordinary people who work hard during the week and who like to relax on weekends by pitting their machines and their nerve against the dunes. Here as elsewhere, though, the contest is an uneven one: when dune meets dune buggy, the dune usually loses. Paradoxically, much of the destruction is inadvertent. Drivers of off-road vehicles do not intend to destroy the vegetation—it just happens. They do not want to hurt the desert, nor do they believe they have. After all, it is just wasteland, is it not? What else is it good for?

Use of off-road vehicles on dunes has increased steadily over the past ten years and shows no sign of leveling off. Between 1964 and 1972, visitor use of the Little Sahara Sand Dunes in central Utah increased tenfold, from 10,000 to

100,000 people yearly. In 1977, visitor-use days there approached 240,000. (A visitor-use day means one visitor for twelve hours.) In 1978, there were over 500,000 visitor-use days on the Algodones Dunes. The heaviest use of the Algodones Dunes occurred in 1980, when an estimated 15,900 off-road vehicles crowded onto the dunes over Thanksgiving weekend. The situation is the same on other desert dune fields that are open to vehicles. Even small dune fields such as the Coral Pink Dunes are visited by 70,000 to 80,000 people yearly, and as many as 1,000 off-road vehicles invade the 2,500 acres of sand at the Coral Pink Dunes on holiday weekends.

Off-road vehicles damage sand dune ecosystems in several ways. First, there is outright destruction of plants by vehicles. Studies have shown that at the Algodones Dunes, areas closed to off-road vehicles support ten times as many plants as open areas. The most popular spots used by off-road vehicles are barren. The Kelso Dunes were a popular spot for off-road vehicles between the early 1950s and 1973, when they were finally closed to vehicular traffic. Aerial photos show that in 1973, the density of shrubs on marginal dunes was half of what it was in 1953. Off-road vehicle clubs staged major events at the Eureka Dunes during the 1960s and 1970s. Even after the dunes were closed to off-road vehicles in 1977, clubs held occasional illegal events. Year after year of such abuse on the Kelso and Eureka dunes reactivated sand on stable marginal dunes as the vegetation was destroyed. Disturbance was so severe that a decade later many spots have not recovered.

Destruction of established plants is only the most visible of the deleterious effects of off-road vehicles. One of the less obvious impacts is compaction of dune sand. As an off-road vehicle passes over an area again and again, grains of sand in its path are pushed together, resulting in decreased pore space between the grains. Decreased pore space means that less water is stored in the sand, and without moisture below the surface, seedlings cannot become established.

Off-road vehicles threaten the survival of endemic plant species on dunes. Endemic dune plants are relatively few in

Off-road vehicles destroy the plant cover on dunes.

number, and their habitat is limited in extent. As individual plants are destroyed, populations diminish, losing an important fraction of their genetic variability. Less variable populations are more likely to succumb than adapt to environmental change, so that seemingly minor destruction of these plants by off-road vehicles could, in fact, cause the extinction of an entire species.

Off-road vehicles have also destroyed the wildness of our desert dune fields. There are few dune fields in the southwestern United States today where one can avoid the ceaseless roar of dune buggies and all-terrain cycles, where interwoven tire tracks do not deface the sand, where acres of sand have not been denuded of plants.

There are other people-related problems on dunes besides those caused by off-road vehicles. Dune fields under the jurisdiction of the National Park Service are popular with sightseers: those at White Sands National Monument have over 500,000 visitors yearly, at Death Valley National Monument, 300,000. The compaction that occurs along trails is probably not a threat to the reproduction and survival of dune plants, but disturbance from foot traffic occasionally results in blowouts on heavily traveled dunes at Death Valley.

Littering and vandalism may be more serious side-effects of foot traffic than disturbance and compaction. Many of the sand pedestals at White Sands have been defaced by grafitti carved into the gypsum sand. Garbage has accumulated around the Little Sahara Sand Dunes wherever people camp, and campers there are gradually denuding the dunes for firewood.

Some dune fields are quarried for quartz sand, as at the dunes near Bouse, Arizona, or for gypsum, as at the Cuatro Cienegas dune field. Quarrying of sand is antithetical to the conservation of dune fields and dune endemics, especially at the Cuatro Cienegas Dunes, where front-end loaders scoop up truckloads of gypsum to be hauled away.

Although the Kelso Dunes are closed to off-road vehicles, cattle range freely over the dune field and graze heavily on many of the grasses and shrubs. Introduction of plants by cattle has already altered the plant community on the Kelso Dunes; continued grazing will only increase invasion by alien plant species.

Of the southwestern dunes managed by the Bureau of Land Management, few are closed altogether to off-road vehicles. Some, such as the Dumont Dunes south of Death Valley, are completely open to motor traffic. Others are partitioned into open and closed sections, with open sections occupying by far the greater area. At the Algodones Dunes, for example, the closed section at the northern end of the dune field is less than twenty percent of the total dune area. Closed areas make up less than one-third of the Little Sahara Sand Dunes. None of the 2,500 acres of sand at the Coral Pink Dunes are closed to off-road vehicles despite the small size of the dune field and the presence of two endemic plant species.

The National Park Service, because of its mandate to preserve and protect our nation's natural and historical heritage, has closed dune fields in its jurisdiction to off-road vehicles. Designation of dune fields as national parks or monuments focuses attention on the dunes. Some of this attention is detrimental, resulting in heavy foot traffic, vandalism, and loss of wildness. Some is beneficial: visitors can be educated to appreciate the fragile dune environment and to protect endemic plants and animals.

Current management of sand dunes is at best an uneasy compromise between management for recreation and management for conservation. Most agencies acknowledge the need to protect irreplaceable endemic plants but also want off-road enthusiasts to enjoy the dunes in their own way. In considering how to balance these antagonistic policies, it is important to remember that off-road vehicles are a fad, yet the damage they do to fragile desert habitats will last centuries after dune buggies and all-terrain cycles have fallen into disuse.

How can dune fields and dune plants in the Southwest best be protected? One possible solution is to place more dune fields under the protection of the National Park Service. This has the advantage of preventing abuse by off-road vehicles and the disadvantage of promoting sightseeing, thus spoiling the wildness of remote dune fields. A better solution would be acquisition of smaller dune fields by conservation groups such as the Nature Conservancy and Defenders of Wildlife. These organizations have already placed hundreds of thousands of acres of endangered habitat beyond the reach of commercial enterprises such as logging, mining, and agriculture. The Nature Conservancy already owns part of at least two dune fields in the West, including 1,300 acres in the sandhills of north-central Nebraska and 226 acres of gypsum sand in western Texas.

A third solution would be to weigh carefully the biological value of every dune field against its recreational value, allowing off-road vehicles on some but not all dune fields. Small dune fields with outstanding biological value should be entirely closed to off-road vehicles and other disruptive uses such as quarrying, grazing, and wood gathering. These dune fields include the Coral Pink, Eureka, Cuatro Ciengas, Monahans, and Kelso dunes. Endemic species and off-road vehicles might be able to coexist on larger dune fields. Under such a plan, vehicles would be limited to certain areas in each dune field. Ideally, more space would be set aside for plants and animals than for vehicles: currently it is the other way around on the Algodones and Little Sahara dunes. The Algodones, Little Sahara, and Gran Desierto dunes are all probably extensive enough for this sort of treatment. Some dune fields, either those with little biological interest or those that have already been denuded by off-road vehicles, could be considered sacrifice areas and opened up completely to off-road vehicles.

When properly protected and preserved, dunes are unique outdoor laboratories for the study of plant ecology and physiology. Many interesting questions about dune plants and their adaptations to a harsh environment remain unanswered. What percentage of seedlings of perennial dune plants actually grows to maturity? How does regional drought affect dune plants? How long does it take for dunes denuded by vehicular traffic to recover?

Sand dunes offer more than projects for biologists. Dunes are home for hundreds of species of plants and animals, many of which can live nowhere else. Their right to exist unharmed should have a greater claim on our sympathies than our right to consume their habitat as if our pleasure were all that mattered, as if we were the axis around which the dune world turns.

From the window of a passing car, dunes look barren and empty. To know the dune country, we must circumvent the four-wheeled barrier that protects and isolates us. Only by walking can we savor the pleasures dune country offers.

Early and late in the day, shadows mold the dunes, giving them shape and substance, boldly outlining the crests and casting the hollows into shade. At noon, the flat sunlight erases the lines that were so sharp, washing out texture and shape. The slopes are steep, and it is a four-limbed scramble to reach the tops of some, but worth it, for on the other side is a hidden world: an expanse of freshly rippled sand or a line of disintegrating animal tracks, a spray of brilliant pink sand verbena or a pinacate beetle trundling on tiptoe across the sand.

Climb to the highest ridge in a dune field and muse on the hills and hollows below. Watch the rising and falling waves of sand and remember that for everything there is a season, and for life on sand dunes it is the season of the wind.

Chapter Notes and Suggested Readings

The Dune Country

Among the many excellent books describing North American deserts are E. C. Jaeger, *The North American deserts* (Stanford, California: Stanford University Press, 1957); F. Shreve and I. L. Wiggins, *Vegetation and flora of the Sonoran Desert* (Stanford, California: Stanford University Press, 1964); R. Kirk, *Desert: The American Southwest* (Boston, Mass.: Houghton Mifflin Company, 1973); and D. E. Brown (ed.), Biotic communities of the American Southwest—United States and Mexico (*Desert Plants* 4:1-132, 1982). A map showing all dune fields in the Southwest can be found in R. S. U. Smith, Sand dunes in the North American deserts, In *Reference handbook on the deserts of North America,* ed. G. L. Bender (Westport, Conn.: Greenwood Press, 1982).

Of Wind and Sand, Sand and Rain

The description of dune formation and morphology is based largely on E. D. McKee, J. R. Douglass, and S. Rittenhouse,

Deformation of lee-side laminae in eolian dunes (*Geological Society of America Bulletin* 82:359-78, 1971); E. D. McKee and R. J. Moiola, Geometry and growth of the White Sands dune field, New Mexico (*Journal of Research of the U.S. Geological Survey* 3:59-66, 1975); E. D. McKee, Structures of dunes at White Sands National Monument (and a comparison with structures of dunes from other selected areas) (*Sedimentology* 7:1-69, 1966); and R. P. Sharp, Kelso Dunes, Mohave Desert, California (*Geological Society of America Bulletin* 77:1045-74, 1966). Less technical discussions can be found in R. Atkinson, *White Sands: Wind, sand and time* (Globe, Arizona: Southwest Parks and Monuments Association, 1977); and in S. A. Trimble, *Great Sand Dunes: The shape of the wind* (Globe, Arizona: Southwest Parks and Monuments Association, 1978). Additional references for most of the topics discussed in this book can be found in J. E. Bowers, The plant ecology of inland dunes in western North America (*Journal of Arid Environments* 5:199-220, 1982).

One of the first scientists to explain the apparent contradiction of moisture in desert dunes was R. A. Bagnold in *The physics of blown sand and desert dunes,* published in 1941 by Methuen in London. Further documentation is contained in R. M. Norris and K. S. Norris, Algodones Dunes of southeastern California (*Geological Society of America Bulletin* 72:605-20, 1961); I. Noy-Meir, Desert ecosystems: Environment and producers (*Annual Review of Ecology and Systematics* 4:25-51, 1973); R. P. Sharp, Kelso Dunes, Mohave Desert, California (*Geological Society of America Bulletin* 77:1045-74, 1966); and in F. Shreve, The sandy areas of the North American desert (*Association of Pacific Coast Geographers Yearbook* 4:11-14, 1938).

Survival

The best discussion of sand movement at the Kelso Dunes is R. P. Sharp, Kelso Dunes, Mohave Desert, California (*Geological Society of America Bulletin* 77:1045-74, 1966). An early description of the peculiar growth of dune buckwheat is that of I. T. Weeks, The life form and habit of *Eriogonum deserticola* Wats. (*Madroño* 1: 260-61, 1929).

Bruce Pavlik investigated the physiology of dune grass in Patterns of water potential and photosynthesis of desert sand dune plants, Eureka Valley, California (*Oecologia* 46:147-54, 1980). Pavlik's physiological studies show that dune grass is a C_4 plant; that is, it fixes (transforms) atmospheric carbon dioxide in the presence of light and the enzyme PEP carboxylase to make the sugars necessary for growth. C_4 plants function best under high temperatures, and the amount of carbon dioxide fixed increases as temperatures rise. Under warm temperatures, plants with the C_4 photosynthetic pathway are able to fix twice as much carbon dioxide as plants using the more common C_3 pathway, and thus C_4 plants can produce more sugars for faster growth. Such productivity is important to C_4 plants on active dunes, where rapid growth is a necessity if plants are to escape burial.

Most plants, including those on dunes, are C_3 plants. They fix carbon dioxide in the presence of light and the enzyme RUDP carboxylase. When temperatures are high and soil moisture is low, plants having C_3 photosynthesis are less productive than those with the C_4 pathway. But with cool temperatures and adequate moisture, C_3 plants outperform C_4 plants. On sand dunes, one would expect C_3 plants such as dune buckwheat to grow fastest in the spring before soil and air temperatures soar above the levels at which the plant can efficiently produce sugars, or in the fall when cooler temperatures prevail once again.

Cacti use a specialized mode of photosynthesis called crassulacean acid metabolism, or CAM for short. CAM enables cacti and other succulent plants to fix carbon dioxide (when the soil is moist enough) at night as well as during the day, unlike non-CAM plants, which can fix carbon dioxide during the day only. All plants take in carbon dioxide through pores, called stomates, on the leaves (or on the stems, in the case of cacti), losing water through their stomates in the process. When temperatures are high enough, non-CAM plants must either suffer a good deal of water loss if they leave their stomates open, or they must lower their rate of photosynthesis, that is, close the stomates. CAM plants, by fixing carbon dioxide at night, lose less water since their stomates are open when the sun is down.

The sand pedestals at White Sands have long interested botanists and plant ecologists. For a more detailed discussion, see F. Emerson, An ecological reconnaissance in the White Sands, New Mexico (*Ecology* 16:226-33, 1935); and L. M. Shields, Gross modifications in certain plant species tolerant of calcium sulfate dunes (*American Midland Naturalist* 50:224-37, 1953). The soil of gravelly desert flats may contain as much as 2,400 parts per million (ppm) total nitrogen, although lower amounts—around 1,800 ppm—are more typical. At White Sands, the lowest nitrogen levels are found on the transverse dunes, where total nitrogen ranges from a low of 91 to a high of 204 ppm. Sand on more stable dunes is higher in nitrogen, containing 181 to 415 ppm, but is still very low when compared with most desert soils. Sand at the Coral Pink Dunes contains 37 to 95 ppm, and that on the Little Sahara Dunes holds only 4 ppm nitrogen.

More information about nutrition of dune plants can be found in L. M. Shields, C. Mitchell, and F. Drouet, Alga- and lichen-stabilized surface crusts as soil nitrogen sources (*American Journal of Botany* 44:489-98, 1957); and in L. H. Wullstein, M. L. Bruening, and W. B. Bollen, Nitrogen fixation associated with sand grain root sheaths (rhizosheaths) of certain xeric grasses (*Physiologia Plantarum* 46:1-4, 1979).

Use of the words "light" and "sunlight" in this discussion is an oversimplification, since plants do not use all of the wavelengths of light present in sunlight. The function of leaf hairs in desert plants is discussed in J. Ehleringer and H. A. Mooney, Leaf hairs: Effect on physiological activity and adaptive value in a desert shrub (*Oecologia* 37:183-200, 1978); and in J. Ehleringer, Leaf absorptances of Mohave and Sonoran desert plants (*Oecologia* 49:366-70, 1981).

Double Bind

Some of the scientists who have investigated reproduction of desert perennials include F. Shreve in Establishment behavior of the palo verde (*Plant World* 14:289-96, 1911); W. C. Sherbrooke in First-year seedling survival of jojoba *(Simmondsia chinensis)* in the Tucson Mountains, Arizona (*Southwestern Naturalist* 22:225-34, 1977); and T. L. Ackerman in Germination and survival of perennial plant species in the Mohave Desert (*Southwestern Naturalist* 24:399-408, 1979).

Little Oaks from Great Acorns

Lake Cabeza de Vaca, and the smaller lakes that followed, are discussed in W. S. Strain, Bolson integration in the Texas–Chihuahua border region, In *Geology of the southern Quitman Mountains area, trans-Pecos Texas, symposium and guidebook 1970* (Midland, Texas: Permian Basin Section, Society of Economic Paleontologists and Mineralogists, 1970); and in C. C. Reeves Jr., Pluvial Lake Palomas, northwestern Chihuahua, Mexico, In *Guidebook of the border region* (New Mexico Geological Society, 20th Field Conference, 1969). The discussion of the growth habits of hoary dicoria is based on B. M. Pavlik, Patterns of water potential and photosynthesis of desert sand dune plants, Eureka Valley, California (*Oecologia* 46:147-54, 1980). Germination of dune plant seeds was investigated by M. A. Maun and S. Riach in Morphology of caryopses, seedlings and seedling emergence of the grass *Calamovilfa longifolia* from various depths of sand (*Oecologia* 49:137-42, 1981). The discussion of seed dormancy is based largely on S. Ellner and A. Shmida, Why are adaptations for long-range seed dispersal rare in desert plants? (*Oecologia* 51:133-44, 1981); and D. L. Venable and L. Lawlor, Delayed germination and dispersal in desert annuals: Escape in space and time (*Oecologia* 46:272-82, 1980). Seed dormancy in annuals is reported in L. Tevis Jr., Germination and growth of ephemerals induced by sprinkling a sandy desert (*Ecology* 39:681-88, 1958). Dispersal of evening primrose seeds is described by L. Tevis Jr. in A population of desert ephemerals germinated by less than one inch of rain (*Ecology* 39:688-95, 1958).

Changes

The discussion of dune field morphology is based on E. D. McKee and R. J. Moiola, Geometry and growth of the White Sands dune field, New Mexico (*Journal of Research of the U.S. Geological Survey* 3:59-66, 1975). Vegetation at White Sands is described in L. M. Shields, Zonation of vegetation within the Tularosa Basin, New Mexico (*Southwestern Naturalist* 1:49-68, 1956). William Reid, University of Texas at El Paso, and his graduate students have studied the vegetation of White Sands more recently.

Succession of ghost town sites in the Mojave Desert is described by R. H. Webb, H. G. Wilshire, and M. A. Henry in Natural recovery of soils and vegetation following human disturbance, In R. H. Webb and H. G. Wilshire (eds.), *Environmental effects of off-road vehicles* (New York: Springer-Verlag, 1983). The discussion of coastal dune vegetation is based on M. L. Kumler, Plant succession on the sand dunes of the Oregon Coast (*Ecology* 50:695-704, 1969). Illustrations of many of the plants can be found in E. J. Larrison, G. W. Patrick, W. H. Baker, and J. A. Yaich, *Washington wildflowers* (Seattle, Washington: Seattle Audubon Society, 1974).

Nowhere Else in the World

The discussion of Cuatro Cienegas is based on E. R. Meyer, Late quaternary palaeocology of the Cuatro Cienegas basin, Coahuila, Mexico (*Ecology* 54:982-95, 1973); B. L. Turner, A new species of *Dyssodia* (Compositae) from north-central Mexico (*Madroño* 21:421-22, 1972); B. L. Turner, Two new gypsophilous species of *Gaillardia* (Compositae) from north-central Mexico (*Southwestern Naturalist* 17:181-90, 1972); B. L. Turner, *Machaeranthera restiformis* (Asteraceae): a bizarre new gypsophile from north-central Mexico (*American Journal of Botany* 60:836-38, 1973); B. L. Turner and A. M. Powell, Deserts, gypsum and endemism, In J. R. Goodin and D. K. Northington (eds.), *Arid land plant resources* (Lubbock, Texas: International Center for Arid and Semi-Arid Land Studies, Texas Tech University, 1979); D. J. Pinkava, Vegetation and flora of the bolson of Cuatro Cienegas region, Coahuila, Mexico: IV. Summary, endemism and corrected catalogue (*Journal of the Arizona-Nevada Academy of Science* 19:23-47, 1984); W. L. Minckley, Cuatro Cienegas fishes: Research review and a local test of diversity versus habitat size (*Journal of the Arizona-Nevada Academy of Science* 19:13-21, 1984); C. J. McCoy, Ecological and zoogeographic relationships of amphibians and reptiles of the Cuatro Cienegas basin (*Journal of the Arizona-Nevada Academy of Science* 19:49-59, 1984); and R. Hershler, The hydrobiid snails (Gastropoda: Rissoacea) of the Cuatro Cienegas basin:Systematic relationships and ecology of a unique fauna (*Journal of the Arizona-Nevada Academy of Science* 19:61-76, 1984).

The Monahans Sandhills and shinnery oak are described in R. Bedichek, *Adventures with a Texas naturalist* (New York: Doubleday, 1947). The discussion of endemics on the Algodones Dunes (and of Peirson's locoweed in particular) owes much to R. C. Barneby, Atlas of North American *Astragalus,* parts I and II (*Memoirs of the New York Botanical Garden* 13:1-1188, 1964). Biologists who have studied sand food include G. Nabhan, *Ammobroma sonorae,* an endangered parasitic plant in extremely arid North America (*Desert Plants* 2:188-96, 1980); and W. P. Armstrong, Sand food: a strange plant of the Algodones Dunes (*Fremontia* 7:3-9, 1980). B. M. Pavlik describes the growth pattern of dune evening primrose in Patterns of water potential and photosynthesis of desert sand dune plants, Eureka Valley, California (*Oecologia* 46:147-54, 1980).

Sea Sands

Salinity levels of 75 to 2,800 ppm were reported for Pacific Coast beaches in the United States in M. G. Barbour and T. M. DeJong, Response of West Coast beach taxa to salt spray, sea-water inundation and soil salinity (*Bulletin of the Torrey Botanical Club* 104:29-34, 1977). Along the Pacific Coast of central Baja California, beach sand averages 21,000 ppm salt, according to A. F. Johnson, A survey of the strand and dune vegetation along the Pacific and southern gulf coasts of Baja California, Mexico (*Journal of Biogeography* 4:83-99, 1977).

A more technical discussion of adaptations of halophytes can be found in M. M. Caldwell, Physiology of desert halophytes, In R. J. Reinhold and W. H. Queen (eds.), *Ecology of halophytes* (New York: Academic Press, 1974). The discussion of seaside sand verbena is based on S. S. Tillet, The maritime species of *Abronia* (*Brittonia* 19:299-327, 1967); and R. C. Wilson, *Abronia:* IV. Acanthocarp dispersibility and its ecological implications for nine species of *Abronia* (*Aliso* 8:493-506, 1976).

The ecology of silverleaf saltbush was investigated by T. M. DeJong and M. G. Barbour in Contributions to the biology of *Atriplex leucophylla,* a C_4 California beach plant (*Bulletin of the Torrey Botanical Club* 106:9-19, 1979). Function of leaf hairs in saltbushes has been studied by L. A. Mozafar and J. R. Goodin in Vesiculated leaf hairs: A mechanism for salt tolerance in *Atriplex halimus* L. (*Plant Physiology* 45:62-65, 1970), among others.

Soil salinity levels of up to 30,000 ppm have been reported from swales in the Death Valley Dunes by C. B. Hunt and L.W. Durrell in Plant ecology of Death Valley (*U.S. Geological Survey Professional Paper 509,* 1966).

The formation of the Coral Pink Dunes is described in H. E. Gregory, Geology and geography of the Zion Park region, Utah and Arizona (*U.S. Geological Survey Professional Paper 220,* 1950).

The floras upon which this chapter is based are those of E. S. Castle, The vegetation and its relationship to the dune soils of the pink sand dunes of Kane County, Utah (M. S. thesis, Brigham Young University, Provo, Utah, 1954); M. Dedecker, Eureka Dunes: Plant list (Unpublished report prepared for the U.S. Department of the Interior, BLM, California Desert District Office, Riverside, California, 1976); R. S. Felger, Vegetation and flora of the Gran Desierto, Sonora, Mexico (*Desert Plants* 2:87-114, 1980); L. M. Shields, Zonation of vegetation within the Tularosa Basin, New Mexico (*Southwestern Naturalist* 1:49-68, 1956); R. F. Thorne, B. A. Prigge, and J. Henrickson, A flora of the higher ranges and the Kelso Dunes of the eastern Mojave Desert in California (*Aliso* 10:71-186, 1981), and WESTEC Services, Survey of sensitive plants of the Algodones Dunes (Unpublished report prepared for the U.S. Department of the Interior, BLM, California Desert District Office, Riverside, California, 1977).

Between Interstate 10 and Highway 62–180 east of El Paso is a chain of largely stable sand hills that extends north into southeastern New Mexico. The source for this sand was most likely sediments deposited in Lake Cabeza de Vaca, which covered much of southeastern New Mexico, western Texas, and northern Chihuahua in the Pleistocene. The sand hills near El Paso are quite stable. Farther north, sand has been blown into locally active dunes. Dominant species on this dune-and-sand-hill system are sand sage, soaptree yucca, giant dropseed, and broom pea.

A more detailed discussion of the plant geography of southwestern sand dunes can be found in J. E. Bowers, Plant geography of southwestern sand dunes (*Desert Plants* 6:31-42, 51-54, 1984).

Other Deserts, Other Dunes

A worldwide survey of sand seas was made by C. S. Breed, S. C. Fryberger, S. Andrews, C. McCauley, F. Lennartz, D. Gebel, and K. Horstman in Regional studies of sand seas using Landsat (ERTS) imagery, In E. D. McKee (ed.), *A study of global sand seas* (*U.S. Geological Survey Professional Paper 1052,* 1979).

Recent studies of Namibian Desert dunes include M. K. Seely and G. N. Louw, First approximation of the effects of rainfall on the ecology and energetics of a Namib Desert dune ecosystem (*Journal of Arid Environments* 3:25-54, 1980); M. D. Robinson and M. K. Seely, Physical and biotic environments of the southern Namib dune ecosystem (*Journal of Arid Environments* 3:183-203, 1980); and M. K. Seely, M. P. de Vos, and G. N. Louw, Fog imbition, satellite fauna and unusual leaf structure in a Namib Desert dune plant, *Trianthema hereroensis* (*South African Journal of Science* 73:169-72, 1977).

The discussion of Kalahari dunes is based on M. J. A. Werger, Vegetation structure in the southern Kalahari (*Journal of Ecology* 66:933-41, 1978); O. A. Leistner, The plant ecology of the southern Kalahari (*Botanical Survey of South Africa, Memoir no. 38,* 1967); and R. C. Buckley, Parallel dunefield ecosystems:Southern Kalahari and central Australia (*Journal of Arid Environments* 4:287-98, 1981).

Sand dunes in Australia were recently investigated by R. C. Buckley in Central Australian sandridges (*Journal of Arid Environments* 4:287-98, 1981). The discussion of An Nafud is based largely on D. F. Vesey-Fitzgerald, The vegetation of central and eastern Arabia (*Journal of Ecology* 45:779-98, 1957).

Use or Abuse

Data on visitor use of southwestern sand dunes was supplied by J. Ross Arnold, Colorado State University; Charles E. Collins Jr., Coral Pink Sand Dunes State Reserve; Donald R.

Harper, White Sands National Monument; Thomas L. Jensen, Bureau of Land Management, House Range Resource Area; Jim Neal, The Nature Conservancy, Texas Field Office; Bob Quesenberry, Death Valley National Monument; Robert Webb, U.S. Geological Survey, Research Project Office, Tucson; Roger D. Zortman, Bureau of Land Management, El Centro Resource Area.

A good discussion of the effects of off-road vehicles on dune plants can be found in R. B. Bury and R. A. Luckenbach, Vehicular recreation in arid land dunes: Biotic responses and management alternatives, In R. H. Webb and H. G. Wilshire (eds.), *Environmental effects of off-road vehicles* (New York: Springer-Verlag, 1983).

Common and Scientific Names of Dune Plants

AUTHOR'S NOTE: A complete scientific plant name has three parts: the genus name, the species name, and the authority, that is, the person who first named and described the plant according to certain long-established rules of botanical nomenclature. If a taxonomist later decides that the species belongs in a different genus, his name becomes part of the authority. Thus *Ambrosia dumosa* (Gray) Payne indicates that Asa Gray described the plant originally, then Willard Payne placed it in the genus *Ambrosia*.

ajo lily *Hesperocallis undulata* Gray
alkali sacaton *Sporobolus airoides* Torr.
armed senna *Cassia armata* Wats.
Barclay's saltbush *Atriplex barclayana* (Benth.) Dietr.
beach morning glory *Convolvulus soldanella* L.
beach silvertop *Glehnia leiocarpa* Math.
beach strawberry *Fragaria chiloensis* (L.) Duchesne
big galleta grass *Hilaria rigida* (Thurb.) Benth.

big-head sedge *Carex macrocephala* Willd.
big sagebrush *Artemisia tridentata* Nutt.
bindweed heliotrope *Heliotropium convolvulaceum* (Nutt.) Gray
bitterbrush *Purshia tridentata* (Pursh) DC.
blackbrush *Coleogyne ramosissima* Torr.
blue palo verde *Cercidium floridum* Benth.
Borrego locoweed *Astragalus lentiginosus* Dougl. var *borreganus* Jones
bracken fern *Pteridium aquilinum* (L.) Kuhn
brittlebush *Encelia farinosa* Gray
broom pea *Psorothamnus scoparius* (Gray) Rydb.
Buckley's penstemon *Penstemon buckleyi* Penn.
California croton *Croton californicus* Muell. Arg.
candelilla *Euphorbia antisyphilitica* Zucc.
cane grass *Zygochloa paradoxa* (R. Br.) S. T. Blake
cardon *Pachycereus pringlei* (S. Wats.) Britt. & Rose
catclaw *Acacia greggii* Gray
cheesebush *Hymenoclea salsola* Torr. & Gray
claret cups hedgehog *Echinocereus triglochidiatus* Engelm.
coastal bursage *Ambrosia chamissonis* Greene
coastal lupine *Lupinus littoralis* Dougl.
cobweb phlox *Eriastrum densifolium* (Benth.) Mason
contorted aster *Machaeranthera restiformis* Turner
creosote bush *Larrea tridentata* (DC.) Cov.
crucifixion thorn *Koeberlinia spinosa* Zucc.
desert holly *Atriplex hymenelytra* (Torr.) Wats.
desert marigold *Baileya pleniradiata* Harv. & Gray
desert saltbush *Atriplex polycarpa* (Torr.) Wats.
desert willow *Chilopsis linearis* (Cav.) Sweet
Drummond's goldenweed *Isocoma drummondii* (Torr. & Gray) Greene
dune buckwheat *Eriogonum deserticola* Wats.
dune caltrop *Neruda procumbens* L.
dune devil's claw *Proboscidea sabulosa* Correll
dune evening primrose *Oenothera avita* W. Klein ssp. *eurekensis* (Munz) Klein

dune grass *Swallenia alexandrae* (Swallen) Soderstrom & Decker
dune peabush *Psorothamnus emoryi* (Gray) Rydb.
dune spectacle-pod *Dimorphocarpa pinnatifida* Rollins
dune sunflower *Helianthus niveus* (Benth.) Brandegee ssp. *tephrodes* (Gray) Heiser
d'Urville's panic grass *Panicum urvilleanum* Kunth
evening primrose *Oenothera deltoides* Torr. & Frem.
fourwing saltbush *Atriplex canescens* (Pursh) Nutt.
frankenia *Frankenia palmeri* Wats.
Fremont dalea *Psorothamnus fremontii* (Torr.) Barneby
Gambel oak *Quercus gambelii* Nutt.
giant cactus *Carnegiea gigantea* (Engelm.) Britt. & Rose
giant dropseed *Sporobolus giganteus* Nash.
giant sand reed *Calamovilfa gigantea* (Nutt.) Scribn. & Merr.
giant Spanish needles *Palafoxia arida* Turner & Morris var. *gigantea* (Jones) Turner & Morris
greasewood *Sarcobatus vermiculatus* (Hook.) Torr.
guayule *Parthenium argentatum* Gray
gypsum blanketflower *Gaillardia gypsophila* Turner
gypsum stinkweed *Dyssodia gypsophila* Turner
hoary dicoria *Dicoria canescens* Torr. & Gray
honey mesquite *Prosopis glandulosa* Torr.
hooded spurge *Euphorbia carunculata* Waterfall
hopsage *Grayia spinosa* (Hook.) Moq.
huckleberry *Vaccinium ovatum* Pursh
Indian rice grass *Oryzopsis hymenoides* (Roem. & Schult.) Rick.
ironwood *Olneya tesota* Gray
Joshua tree *Yucca brevifolia* Engelm.
juniper *Juniperus osteosperma* (Torr.) Little
kinnikinnik *Arctostaphylos uva-ursi* (L.) Spreng.
lechuguilla *Agave lecheguilla* Torr.
little-leaf palo verde *Cercidium microphyllum* (Torr.) Rose & Johnston

lodgepole pine *Pinus contorta* Dougl.
long-leaf sunflower *Wyethia scabra* Hook. var. *attenuata* W. A. Weber
lupine *Lupinus sparsiflorus* Benth.
marram grass *Ammophila arenaria* (L.) Link.
marriola *Parthenium incanum* H. B. K.
mesa dropseed *Sporobolus flexuosus* (Thurb.) Rydb.
mistletoe *Phoradendron* spp.
Mojave sage *Salvia mohavensis* Greene
Mojave yucca *Yucca schidigera* Roezl.
moon pod *Selinocarpus purpusianus* Heimerl
Mormon tea *Ephedra* spp.
narrow-leaved sand verbena *Abronia angustifolia* Greene
needle-and-thread grass *Stipa* spp.
ocotillo *Fouquieria splendens* Engelm.
palafoxia *Palafoxia sphacelata* (Torr.) Gray
pale evening primrose *Oenothera pallida* Lindl.
Palmer's brittlebush *Encelia palmeri* Vasey & Rose
Parry's spurge *Euphorbia parryi* Engelm.
Peirson's locoweed *Astragalus magdalenae* Greene var. *peirsonii* (Munz & Barneby) Barneby
pickleweed *Allenrolfea occidentalis* (Wats.) Kuntze
plains penstemon *Penstemon ambiguus* Torr.
plains prickly pear *Opuntia macrorhiza* Engelm.
plains yucca *Yucca campestris* McKelvey
pleated coldenia *Tiquilia plicata* (Torr.) Richards.
ponderosa pine *Pinus ponderosa* Lawson
popcorn flower *Cryptantha* spp.
prince's plume *Stanleya pinnata* (Pursh) Britton
purple three-awn *Aristida purpurea* Nutt.
red brome *Bromus rubens* L.
rhododendron *Rhododendron macrophyllum* G. Don.
Riddell's groundsel *Senecio riddellii* Torr. & Gray
ringwing *Cycloloma atriplicifolium* (Spreng.) Coult.
Rio Grande cottonwood *Populus fremontii* Wats. var. *wislizenii* (Torr.) Wats.

rosemary mint *Poliomintha incana* (Torr.) Gray
rubber rabbitbrush *Chrysothamnus nauseosus* (Pall.) Britton
Russian thistle *Salsola kali* L.
saguaro *Carnegiea gigantea* (Engelm.) Britt. & Rose
salt grass *Distichlis stricta* (Torr.) Rydb.
sand beargrass *Nolina arenicola* Correll
sand dropseed *Sporobolus cryptandrus* (Torr.) Gray
sand food *Ammobroma sonorae* Torr.
sand locoweed *Astragalus lentiginosus* Dougl. var. *micans* Barneby
sand sage *Artemisia filifolia* Torr.
sand verbena *Abronia villosa* Wats.
scurf-pea *Psoralea lanceolata* Pursh
seaside sand verbena *Abronia maritima* Nutt.
seepweed *Suaeda* spp.
shadscale *Atriplex confertifolia* (Torr. & Frem.) Wats.
shallon *Gaultheria shallon* Pursh
shinnery oak *Quercus havardii* Rydb.
silver spurge *Euphorbia leucophylla* Benth.
silverleaf saltbush *Atriplex leucophylla* (Moq.) D. Dietr.
silvery locoweed *Astragalus magdalenae* Greene var. *niveus* (Rydb.) Barneby
silvery sophora *Sophora stenophylla* Gray
silvery sunflower *Helianthus niveus* (Benth.) Brandegee ssp. *niveus*
Sitka spruce *Picea sitchensis* (Bong.) Carr.
skeleton weed *Eriogonum deflexum* Torr.
soaptree yucca *Yucca elata* Engelm.
sotol *Dasylirion palmeri* Trel.
Spanish needles *Palafoxia arida* Turner & Morris var. *arida*
spectacle-pod *Dithyrea wislizenii* Engelm.
squawbush *Rhus trilobata* Nutt.
stick-leaf *Petalonyx crenatus* Gray
strawberry hedgehog *Echinocereus enneacanthus* Engelm.

tarbush *Flourensia cernua* DC.
Texas croton *Croton texensis* (Klotsch) Muell. Arg.
thickspike wheatgrass *Agropyron dasystachyum* (Hook.) Scribn.
thorny dalea *Psorothamnus polyadenius* (Torr.) Rydb.
Thurber's penstemon *Penstemon thurberi* Torr.
Thurber's stick-leaf *Petalonyx thurberi* Gray
triangle-leaf bursage *Ambrosia deltoidea* (Torr.) Payne
varilla *Varilla mexicana* Gray
Welsh's milkweed *Asclepias welshii* Holmgren & Holmgren
white bursage *Ambrosia dumosa* (Gray) Payne
whitethorn *Acacia vernicosa* Standl.
Wiggins' croton *Croton wigginsii* Wheeler
wild buckwheat *Eriogonum* spp.
wild heliotrope *Phacelia* spp.
wolfberry *Lycium* spp.
yellow pine *Pinus ponderosa* Lawson
yellow sand verbena *Abronia latifolia* Eschsch.

A Selective Guide to Southwestern Dunes

This is not a comprehensive guide to dune fields in the Southwest; rather, it includes those I am familiar with and omits those I have not seen or for which there is little published information.

I have suggested additional readings, including wildflower guides, for every dune field in this section. Please keep in mind that many of the plants found on sand dunes are rare, and some are in danger of extinction.

Map coverage refers, for the United States, to the 1:250,000 maps of the western United States, published by the U.S. Geological Survey; and for Mexico, to the 1:250,000 Americas series, published by the Commission on Cartography, Pan American Institute of Geography and History. The maps of the western United States are widely available in map stores, backpacking stores, and public and university libraries. They can also be ordered directly from: U.S. Geological Survey, Western Distribution Branch, P.O. Box 25286, Federal Center, Denver, CO 80225.

Algodones Dunes

Area: 200 square miles
Elevation: 300 feet
Highest dunes: 300 feet
Morphology: The Algodones Dunes are complex dunes. Basically, the dune field is dominated by enormous transverse dunes that appear to be a mass of peaks and valleys. Barchan dunes are common on swales at the southern end of the dune field. Sand source for the dune field was sediments deposited in Lake Cahuilla, a much larger forerunner of the Salton Sea.
Vegetation: The stable marginal dunes support many species of shrubs and, in the spring, wildflowers. Dune peabush, creosote bush, white bursage, and Mormon tea are the most common shrubs. Sand food may be found on the less active dunes in May. Dune buckwheat, dune sunflower, giant Spanish needles, Peirson's locoweed, and d'Urville's panic grass, all dune endemics, grow on active dunes in the center of the dune field.
Access: The Algodones Dunes are open to off-road vehicles south of Highway 78, which crosses the dune field from east to west. The dune field is closed to vehicles north of Highway 78. There is a developed campground south of Highway 78 on the west side of the dune field, and many informal campgrounds on the flats east of the dunes. There is additional access to the dune field at the southern end, where Interstate 8 crosses the dunes. The dune field is most crowded on weekends during the fall and winter, particularly on holiday weekends.
Map Coverage: El Centro, California–Arizona 1:250,000
Location: Imperial County, California, west of Yuma, Arizona
Further Reading:

Dodge, N. N. 1951. *Flowers of the Southwest deserts.* Globe, Arizona: Southwest Parks and Monuments Association.

Munz, P. A. 1962. *California desert wildflowers.* Berkeley: University of California Press.

Norris, R. M., and K. S. Norris. 1961. Algodones Dunes of southeastern California. *Geological Society of America Bulletin* 72:605-20.

Coral Pink Dunes

Size: 7 square miles
Elevation: 6,000 feet
Highest dunes: 75 feet
Morphology: Stable parabolic dunes line the margins of the dune field. The active dunes in the center of the dune field are transverse and barchan dunes. Sand source for the dune field is weathered Navajo sandstone, which forms many of the sandstone ramps and cliffs in the area.
Vegetation: Notice the stragglers on the stable, marginal dunes: Gambel oak, big sagebrush, bitterbrush, rubber rabbitbrush, and others. Long-leaf sunflower is the most common plant on the active dunes; its mounds are conspicuous even in winter after the plants have lost their leaves. Look for scurfpea, sand sage, and giant sand reed on active and stable dunes. In the summer the endemic Welsh's milkweed may be found on dunes and in swales. Notice both the living and dead ponderosa pines scattered over the dune field.
Access: The Coral Pink Dunes are open to off-road vehicles. There is a campground run by the Utah State Parks Board at the dune field. Summers are the busy season; the best time to visit the Coral Pink Dunes is before May or after September.
Map Coverage: Cedar City, Utah 1:250,000
Location: West of Kanab in Kane County, Utah
Further Reading:

Gregory, H. E. 1950. Geology and geography of the Zion Park Region, Utah and Arizona. *U.S. Geological Survey Professional Paper 220.*

Krutch, J. W. 1968. Coral dunes. *Audubon* 70:66-79.

Nelson, R. A. 1976. *Plants of Zion National Park.* Springdale, Utah: Zion Natural History Association.

Cuatro Cienegas Dunes

Area: 8 square miles
Elevation: 2,300 feet
Highest dunes: 40 feet
Morphology: The dune field is a mosaic of active dunes, stable dunes, flat beds of gypsum, and sand dune pedestals, all interacting in a cycle of dune formation, stabilization, and reactivation. As gypsum pedestals slowly erode, they supply loose sand, which is blown into active dunes and eventually anchored by plants. Some of the stable dunes harden into gypsum pedestals, which finally erode, continuing the cycle. Sand source for the dunes is gypsum sediments deposited in Laguna del Churince, east of the dune field.
Vegetation: Honey mesquite is common on hummocks throughout the dune field. A yucca, *Yucca treculeana,* and a sotol, *Dasylirion palmeri,* are common on stable marginal dunes. Other common shrubs on the dunes are moon pod, stick-leaf, varilla, and Drummond's goldenweed. The mound-shaped cactus with stiff, curving spines is strawberry hedgehog. Look for contorted aster, which is endemic to the Cuatro Cienegas dune field, on active and stable dunes. Gypsum blanketflower and gypsum stinkweed, which are also endemic, are less common than contorted aster, and may be found on stable dunes near the edge of the dune field.
Access: A tourist visa and a car permit must be presented when crossing the border into Mexico—standard border procedures. The dune field lies west of Highway 30; several ungraded but passable dirt roads lead from the highway to the dunes. Informal camping is permitted on or near the dunes. There are no developed campgrounds.
Map Coverage: Tlahualilo de Zaragoza, Mexico 1:250,000

Location: Coahuila, Mexico, about 120 miles southwest of Monclova

Further Reading:

Franz, C. 1972. *The people's guide to Mexico.* Santa Fe, New Mexico: J. Muir Publications.

Pinkava, D. J. 1984. Vegetation and flora of the bolson of Cuatro Cienegas region, Coahuila, Mexico: IV. Summary, endemism and corrected catalogue. *Journal of the Arizona-Nevada Academy of Science* 19:23-47.

Death Valley Dunes

Area: 31 square miles
Elevation: sea level
Highest dunes: 80 feet
Morphology: Transverse dunes make up five-sixths of the dune field; the remainder is a sandy sheet. Although the tallest dunes near the south end of the dune field are active, the dune field itself is not moving up or down the valley.
Vegetation: The central dunes are barren or nearly so. On hard, silty swales between the central dunes, pickleweed, creosote bush, and fourwing saltbush may be found. Look carefully on steep slopes for pads of creosote leaves and flowers where the shrubs were buried by advancing dunes. On the south side of the dune field many marginal dunes have been stabilized by honey mesquite and creosote bush. On the east side of the dune field near the picnic area, look for sand dropseed, alkali sacaton, and seepweed, as well as honey mesquite and pickleweed.
Access: The Death Valley Dunes are closed to off-road vehicles. Camping is not permitted at the dune field, but there is camping nearby at Stovepipe Wells. The Death Valley National Monument visitor center has information on other campgrounds in the monument. The tallest dunes lie within easy walking distance of Highway 190. Rangers lead scheduled tours of the dune field; the visitor center posts dates and times.

Death Valley can be quite crowded during the winter, particularly on holiday weekends.

Map Coverage: Death Valley, California–Nevada 1:250,000

Location: Death Valley National Monument, Inyo County, California

Further Reading:

Bessken, D. P. 1980. *Sand dune story.* Death Valley, California: Death Valley Natural History Association.

Ferris, R. S. 1974. *Death Valley wildflowers.* Death Valley, California: Death Valley Natural History Association.

Hunt, C. B., and L. W. Durrell. 1966. Plant ecology of Death Valley. *U.S. Geological Survey Professional Paper 509.*

Jaeger, E. C. 1940. *Desert wild flowers.* Stanford, California: Stanford University Press.

Munz, P. A. 1962. *California desert wildflowers.* Berkeley: University of California Press.

Norris, L. L. 1982. *A checklist of the vascular plants of Death Valley National Monument.* Death Valley, California: Death Valley Natural History Association.

Eureka Dunes

Area: 3 square miles

Elevation: 3,000 feet

Highest dunes: 660 feet

Morphology: The Eureka Dunes have been classified as a "sand mountain" by one geologist. They are quite active, but do not appear to be advancing toward the Last Chance Range, which lies behind them. Sand source for the dune field was sediments deposited in a Pleistocene lake in the Eureka Valley.

Vegetation: On the northern and eastern flanks of the dune field, hummocks have been stabilized by thorny dalea. Scattered on the sand between the hummocks are pleated coldenia and dune locoweed. Higher on the dune field, in fact nearly to the top, are hummocks of dune grass. In the spring, if winter rains have been good, seedlings of dune grass may be seen scattered across the gentler slopes. Near the playa, on the west side of the dune field, look for three different species of saltbush: desert saltbush, shadscale, and fourwing saltbush. Indian rice grass is common on the sandy plain on the east side of the dune field. In the summer hoary dicoria may be abundant on the dunes. Three species of plants are endemic to the Eureka Dunes: dune grass, dune locoweed, and dune evening primrose. Look for dune grass at any time of year, for the other two in the spring.

Access: The Eureka Dunes are closed to all vehicles and are patrolled by Bureau of Land Management rangers to enforce closure. Camping is permitted on and near the dunes; there are no developed campgrounds at the dune field. A good dirt road leads to the edge of the dunes from the Saline Valley road.

Map Coverage: Goldfield, Nevada–California 1:250,000

Location: Inyo County, California, at the south end of the Eureka Valley

Further Reading:

Dedecker, M. 1976. The Eureka Dunes. *Fremontia* 4:17-20.

Henry, M. A. 1979. A rare grass on the Eureka Dunes. *Fremontia* 7:3-6.

Jaeger, E. C. 1940. *Desert wild flowers.* Stanford, California: Stanford University Press.

Munz, P. A. 1962. *California desert wildflowers.* Berkeley: University of California Press.

Gran Desierto Dunes

Area: 1,700 square miles

Elevation: 500 feet

Highest dunes: 600 feet

Morphology: The tallest dunes in the Gran Desierto dune field are star dunes, which usually have three arms radiating from the central peak. Most of the lower dunes are barchans or longitudinal dunes. Here, more than in any other southwestern dune field, one gets the impression that individual dunes are pinned around the base by vegetation, while the barren crests move back and forth in the wind.

Vegetation: The most common shrubs on the Gran Desierto Dunes are Mormon tea, dune buckwheat, Wiggins' croton, and white bursage. Notice how the dune-adapted endemics—Wiggins' croton and dune buckwheat—grow on the more active dune slopes, whereas Mormon tea and white bursage are restricted to the stabler slopes. If winter rains have been good, one should be able to find many different species of wildflowers in the swales. Look for sand verbena, evening primrose, lupine, wild heliotrope, ajo lily, and popcorn flower in the early spring. Sand food may be found on stable dunes in May.

Access: A visa or a car permit is not necessary to visit the north end of the Gran Desierto dune field, but both are required to travel south of Sonoyta. The Gran Desierto dunes are open to vehicles. Four-wheel drive is necessary to reach some parts of the dune field because the roads are sandy and very soft in places. Marginal dunes are accessible by passenger car over ungraded dirt roads. Informal camping is permitted on and near the dunes. The closest developed campground is at Organ Pipe Cactus National Monument in Arizona.

Map Coverage: Mexicali, Mexico 1:250,000

Location: South of Highway 2, about 75 miles west of Sonoyta and 40 miles east of San Luis, in Sonora, Mexico

Further Reading:

Dodge, N. N. 1969. *Flowers of the Southwest deserts.* Globe, Arizona: Southwest Parks and Monuments Association.

Felger, R. S. 1980. Vegetation and flora of the Gran Desierto, Mexico. *Desert Plants* 2:87-114.

Franz, C. 1972. *The people's guide to Mexico.* Santa Fe, New Mexico: J. Muir Publications.

Kelso Dunes

Area: 45 square miles
Elevation: 2,000 feet
Highest dunes: 550 feet
Morphology: There are four major sand ridges in the dune field. Superimposed on these are active transverse dunes. Stable parabolic dunes lie along the southern and eastern sides of the dune field.
Vegetation: The most abundant shrubs on the stable dunes are creosote bush and white bursage. Desert willow trees are scattered over the unstable transverse dunes. Three different species of grass grow on the Kelso Dunes: big galleta grass is most abundant on the lower, stable parts of the dune field; d'Urville's panic grass is most abundant on the highest, active slopes; and Indian rice grass is scattered here and there on stable and active dunes. In the spring the endemic Borrego locoweed with its bright, red-violet flowers is common on active dunes. Notice the hummocks of California croton in spring and summer; in the winter the stems die back to the ground, and winds tend to disperse the hummocks.
Access: The Kelso Dunes are closed to all vehicles and are patrolled by Bureau of Land Management rangers to enforce closure. There is a good dirt road to the south side of the dune field. Informal camping is permitted on or near the dunes; the closest developed campground is at Providence Mountains State Recreation Area.
Map Coverage: Needles, California–Arizona 1:250,000
Location: South of Baker, between Interstate 15 and Interstate 40, in San Bernardino County, California
Further Reading:

Jaeger, E. C. 1940. *Desert wildflowers.* Stanford, California: Stanford University Press.

Munz, P. A. 1962. *California desert wildflowers.* Berkeley: University of California Press.

Sharp, R. P. 1966. Kelso Dunes, Mohave Desert, California. *Geological Society of America Bulletin* 72:605-20.

Thorne, R. F., B. A. Prigge, and J. Henrickson. 1981. A flora of the higher ranges and the Kelso Dunes of the eastern Mojave Desert in California. *Aliso* 10:71-186.

Monahans Sandhills

Area: 6 square miles
Elevation: 2,800 feet
Highest dunes: 50 feet
Morphology: Monahans Sandhills lie at the southern end of a dune system that stretches into eastern New Mexico. Two different sand formations constitute the dune field. The older formation, which once formed extensive dunes throughout the area, is reddish brown due to iron oxide stains on the sand grains. The younger formation, which is cream colored, lies on top of the older formation and has been blown into active transverse and barchan dunes.
Vegetation: Stable dunes support a dwarfed woodland of shinnery oak and honey mesquite; sand sage may be found on more active dunes. The summer flowering season can be very colorful: bindweed heliotrope, palafoxia, pale evening primrose, and spectacle-pod are some of the common sand dune wildflowers.
Access: Dunes in Monahans Sandhills State Park are closed to off-road vehicles. There is a campground at the dune field.
Map Coverage: Pecos, Texas 1:250,000
Location: Monahans Sandhills State Park, southwest of Odessa, in Ward and Winkler counties, Texas
Further Reading:

Eifler, G. H. Jr. 1970. Monahans Sandhills State Park. In *Geologic and historic guide to the state parks of Texas, Guidebook 10,* ed. R. A. Maxwell. Austin: Bureau of Economic Geology, University of Texas.

Warnock, B. H. 1974. *Wildflowers of the Guadalupe Mountains and the sand dune country, Texas.* Alpine, Texas: Sul Ross State University.

Samalayuca Dunes

Area: 100 square miles
Elevation: 4,300 feet
Highest dunes: 550 feet
Morphology: One geologist has said that the Samalayuca Dunes are "non-descript piles of sand with irregular and inconsistent shapes." The tall dunes are reminiscent of the Algodones dune field, where smaller dunes are carved upon larger ones. Sand source for the dune field was sediments in Lake Palomas—a Pleistocene lake that filled the basin west of the dune field—and its precursor, Lake Cabeza de Vaca, which covered much of Chihuahua, western Texas, and southeastern New Mexico.
Vegetation: Stable dunes on the margins of the dune field support honey mesquite, broom pea, fourwing saltbush, creosote bush, and many other species. On active dunes are hummocks of sand sage, rosemary mint, and soaptree yucca. After summer rains, many species of wildflowers, including bindweed heliotrope, spectacle-pod, plains penstemon, pale evening primrose, Texas croton, narrow-leaved sand verbena, desert marigold, and palafoxia are likely to be in bloom. Common grasses are giant dropseed and mesa dropseed.
Access: Although the Samalayuca Dunes are not closed to off-road vehicles, they have not attracted many off-road enthusiasts. The tallest dunes, which lie east of Highway 45, are accessible by an ungraded dirt road that leaves the highway about two miles north of the little town of Estacion Samalayuca. Highway 45 runs right through the dune field, and there are turnouts here and there. Informal camping is permitted on or near the dune field; there are no developed campgrounds. A tourist visa and car permit are required for crossing the border.
Map Coverage: Ciudad Juarez, Mexico 1:250,000
Location: Along Highway 45 about 30 miles south of Ciudad Juarez, in Chihuahua, Mexico

Further Reading:
Franz, C. 1972. *The people's guide to Mexico.* Santa Fe, New Mexico: J. Muir Publications.
Warnock, B. H. 1974. *Wildflowers of the Guadalupe Mountains and the sand dune country, Texas.* Alpine, Texas: Sul Ross State University.
Webb, D. S. 1969. Facets of the geology of the Sierra del Presidio area, north-central Chihuahua. In *Guidebook of the Border Region, 20th Field Conference.*

White Sands

Area: 275 square miles
Elevation: 4,000 feet
Highest dunes: 40 feet
Morphology: Sand source for the dune field is gypsum sediments deposited in Lake Lucero at the southwestern end of the dune field. Barren, dome-shaped dunes immediately east of Lake Lucero give way to active transverse dunes to the northeast and east. The transverse dunes, which make up much of the dune field, are bounded on the northeast and east by stable parabolic dunes.
Vegetation: On transverse dunes one will see squawbush, Rio Grande cottonwood, fourwing saltbush, soaptree yucca, rosemary mint, and Mormon tea. Notice how large these plants are, particularly squawbush and saltbush. Stable parabolic dunes support most of these species and many others. The only plants in the Heart of the Dunes grow on swales between the barren transverse dunes: alkali sacaton, Indian rice grass, narrow-leaved sand verbena, and pale evening primrose.
Access: White Sands dune field lies partly within White Sands National Monument and partly within White Sands Missile Range. The missile range is closed to the public. In the monument, the dunes are closed to off-road vehicles. A paved road goes from the visitor center to the heart of the dune field. Camping is not permitted at the monument, but there are

forest service campgrounds in the Sacramento Mountains to the east. Rangers lead scheduled tours of the dunes; the visitor center posts dates and times.

Map Coverage: Las Cruces, New Mexico–Texas 1:250,000

Location: White Sands National Monument, southwest of Alamogordo, in Dona Ana and Otero counties, New Mexico

Further Reading:

Atkinson, R. 1977. *White Sands: Wind, sand and time.* Globe, Arizona: Southwest Parks and Monuments Association.

Dodge, N. N. 1969. *Flowers of the Southwest deserts.* Globe, Arizona: Southwest Parks and Monuments Association.

McKee, E. D., and J. R. Douglass. 1971. Growth and movement of dunes at White Sands National Monument, New Mexico. *U.S. Geological Survey Professional Paper 750-D.*

Patraw, P. M. 1970. *Flowers of the Southwest mesas.* Globe, Arizona: Southwest Parks and Monuments Association.

Shields, L. M. 1956. Zonation of vegetation within the Tularosa Basin, New Mexico. *Southwestern Naturalist* 1:49-68.

Index to Plant Names

General Index

About the Author

Janice Emily Bowers is the author of *Fear Falls Away, A Full Life in a Small Place, The Mountains Next Door,* and other books about the natural world. Since 1982 she has worked as a botanist for the U.S. Geological Survey. She lives on the outskirts of Tucson with her hushand, Steve McLaughlin, her cat, Katie, and a red-spotted toad, Little Buddy.